Journal of Applied Logics - IfCoLog Journal of Logics and their Applications

Volume 10, Number 2

April 2023

Disclaimer

Statements of fact and opinion in the articles in Journal of Applied Logics - IfCoLog Journal of Logics and their Applications (JALs-FLAP) are those of the respective authors and contributors and not of the JALs-FLAP. Neither College Publications nor the JALs-FLAP make any representation, express or implied, in respect of the accuracy of the material in this journal and cannot accept any legal responsibility or liability for any errors or omissions that may be made. The reader should make his/her own evaluation as to the appropriateness or otherwise of any experimental technique described.

© Individual authors and College Publications 2023
All rights reserved.

ISBN 978-1-84890-431-6
ISSN (E) 2631-9829
ISSN (P) 2631-9810

College Publications
Scientific Director: Dov Gabbay
Managing Director: Jane Spurr

http://www.collegepublications.co.uk

All rights reserved. No part of this publication may be reproduced, stored in a retrieval system or transmitted in any form, or by any means, electronic, mechanical, photocopying, recording or otherwise without prior permission, in writing, from the publisher.

Editorial Board

Editors-in-Chief
Dov M. Gabbay and Jörg Siekmann

Marcello D'Agostino	Melvin Fitting	Henri Prade
Natasha Alechina	Michael Gabbay	David Pym
Sandra Alves	Murdoch Gabbay	Ruy de Queiroz
Arnon Avron	Thomas F. Gordon	Ram Ramanujam
Jan Broersen	Wesley H. Holliday	Chrtian Retoré
Martin Caminada	Sara Kalvala	Ulrike Sattler
Balder ten Cate	Shalom Lappin	Jörg Siekmann
Agata Ciabattoni	Beishui Liao	Marija Slavkovik
Robin Cooper	David Makinson	Jane Spurr
Luis Farinas del Cerro	Réka Markovich	Kaile Su
Esther David	George Metcalfe	Leon van der Torre
Didier Dubois	Claudia Nalon	Yde Venema
PM Dung	Valeria de Paiva	Rineke Verbrugge
David Fernandez Duque	Jeff Paris	Heinrich Wansing
Jan van Eijck	David Pearce	Jef Wijsen
Marcelo Falappa	Pavlos Peppas	John Woods
Amy Felty	Brigitte Pientka	Michael Wooldridge
Eduaro Fermé	Elaine Pimentel	Anna Zamansky

SCOPE AND SUBMISSIONS

This journal considers submission in all areas of pure and applied logic, including:

pure logical systems
proof theory
constructive logic
categorical logic
modal and temporal logic
model theory
recursion theory
type theory
nominal theory
nonclassical logics
nonmonotonic logic
numerical and uncertainty reasoning
logic and AI
foundations of logic programming
belief change/revision
systems of knowledge and belief
logics and semantics of programming
specification and verification
agent theory
databases

dynamic logic
quantum logic
algebraic logic
logic and cognition
probabilistic logic
logic and networks
neuro-logical systems
complexity
argumentation theory
logic and computation
logic and language
logic engineering
knowledge-based systems
automated reasoning
knowledge representation
logic in hardware and VLSI
natural language
concurrent computation
planning

This journal will also consider papers on the application of logic in other subject areas: philosophy, cognitive science, physics etc. provided they have some formal content.

Submissions should be sent to Jane Spurr (jane@janespurr.net) as a pdf file, preferably compiled in LaTeX using the IFCoLog class file.

CONTENTS

ARTICLES

Joining Formal and Cognitive Reasoning 115
   Christoph Beierle, Marco Ragni, Frieder Stolzenburg and Matthias Thimm

The Weak Completion Semantics and Counterexamples 121
   Meghna Bhadra and Steffen Hölldobler

Epistemic State Mappings among Ranking Functions and Total Preorders . 155
   Jonas Philipp Haldimann, Christoph Beierle and Gabriele Kern-Isberner

Probabilistic Deontic Logics for Reasoning about Uncertain Norms 193
   Vincent de Wit, Dragan Doder and John Jules Meyer

Activation-based Conditional Inference 221
   Marco Wilhelm, Diana Howey, Gabriele Kern-Isberner, Kai Sauerwald and Christoph Beierle

Do Humans Find Postulates of Belief Change Plausible? 249
   Clayton Kevin Baker and Thomas Meyer

Joining Formal and Cognitive Reasoning

Christoph Beierle
FernUniversität in Hagen, Hagen, Germany
`christoph.beierle@fernuni-hagen.de`

Marco Ragni
Technische Universität Chemnitz, Chemnitz, Germany
`marco.ragni@hsw.tu-chemnitz.de`

Frieder Stolzenburg
Hochschule Harz, Wernigerode, Hagen, Germany
`fstolzenburg@hs-harz.de`

Matthias Thimm
FernUniversität in Hagen, Hagen, Germany
`matthias.thimm@fernuni-hagen.de`

Because information for real life AI applications is usually pervaded by uncertainty and subject to change, nonclassical reasoning approaches are required. At the same time, psychological findings indicate that human reasoning cannot be completely described by classical logical systems. Sources of explanations are incomplete knowledge, incorrect beliefs, or inconsistencies. A wide range of reasoning mechanism has to be considered, such as analogical or defeasible reasoning, possibly in combination with machine learning methods. The field of knowledge representation and reasoning offers a rich palette of methods for uncertain reasoning both to describe human reasoning and to model AI approaches.

This special issue includes revised and extended versions of the best papers presented at the *7th Workshop on Formal and Cognitive Reasoning (FCR-2021)* which was co-located with the 44th German Conference on Artificial Intelligence (KI-2021). Additionally, the special issue contains further contributions resulting from an open call for papers dedicated to the themes of the workshop. The KI-2021 conference and all its workshops were expected to take place in Berlin, Germany. However, because of the Corona pandemic all were turned into fully virtual events.

As the previous editions of this workshop series, FCR-2021 was organized jointly by two special interest groups of the German Informatics Society (*GI, Gesellschaft für Informatik*), the Special Interest Group on Knowledge Representation and Reasoning (*GI-Fachgrupppe Wissensrepräsentation und Schließen*), and the Special Interest Group on Cognition (*GI-Fachgrupppe Kognition*). FCR-2021 was the 7th Workshop on Formal and Cognitive Reasoning, following previous workshops in Dresden, Germany (2015), Bremen, Germany (2016), Dortmund, Germany (2017), Berlin, Germany (2018), Kassel, Germany (2019), and Bamberg, Germany/online (2020).

The aim of this series of workshops is to address recent challenges and to present novel approaches to uncertain reasoning and belief change in their broad senses, and in particular provide a forum for research work linking different paradigms of reasoning. In 2021, we welcomed especially contributions on intersections between human and formal aspects such as computational thinking. A special focus is on papers that provide a base for connecting formal-logical models of knowledge representation and cognitive models of reasoning and learning, addressing formal and experimental or heuristic issues.

In their paper *The Weak Completion Semantics and Counterexamples*, Meghna Bhadra and Steffen Hölldobler deal with the observations which conclusions participants in an experiment made if the antecedent of a conditional sentence is denied. While most participants concluded that the negation of the consequent holds, a significant number of participants answered *nothing follows* if the antecedent of the conditional sentence was non-necessary. The authors extend the Weak Completion Semantics which correctly models the answers of the majority, but cannot explain the number of *nothing follows* answers, by counterexamples, allowing it to explain the experimental findings.

The paper *Epistemic State Mappings among Ranking Functions and Total Preorders* by Jonas Philipp Haldimann, Christoph Beierle, and Gabriele Kern-Isberner addresses two common models for epistemic states that can represent conditional beliefs, namely ranking functions and total preorders on possible worlds. Connections within and between these frameworks are formalized as *epistemic state mappings*. Postulates concerning the preservation of properties under the application of these mappings are introduced, characterizing important properties like syntax splitting or the compatibility with operations like marginalization and conditionalization. The interrelationships among the postulates for epistemic state mappings within and across the two frameworks are evaluated, establishing dependencies as well as incompatibilities among postulates.

In the contribution *Probabilistic Deontic Logics for Reasoning about Uncertain Norms*, Vincent de Wit, Dragan Doder, and John Jules Meyer present a proof-theoretical and model-theoretical approach to probabilistic logic for reasoning about uncertainty about normative statements. They introduce two logics that extend both the language of monadic deontic logic and the language of probabilistic logic. The first logic allows statements like "the probability that one is obliged to be quiet is at least 0.9". The second logic allows iteration of probabilities in the language. Both logics are axiomatized, the corresponding semantics are provided, and the axiomatizations are shown to be sound and complete. In addition, it is proven that both logics are decidable and that the problem of deciding satisfiability for the simpler of the two logics is in PSPACE, no worse than that of deontic logic.

In the paper *Activation-Based Conditional Inference*, Marco Wilhelm, Diana Howey, Gabriele Kern-Isberner, Kai Sauerwald, and Christoph Beierle develop activation-based conditional inference which combines conditional reasoning and ACT-R, a cognitive architecture developed to formalize human reasoning. The basic idea of activation-based conditional inference is to determine a reasonable, cognitively adequate subset of a conditional belief base before drawing inductive inferences. Central to activation-based conditional inference is the activation function which assigns to the conditionals in the belief base a degree of activation mainly based on the conditional's relevance for the current query and its usage history. The paper presents a blueprint for activation-based conditional inference and illustrates how focusing, forgetting, and remembering are included within the framework.

In the final contribution of this special issue, *Do humans find postulates of belief change plausible?*, Clayton Kevin Baker and Thomas Meyer use various empirical methods to test whether humans agree with postulates of non-monotonic reasoning and belief change. The paper investigates whether postulates of revision and update are plausible with human reasoners when presented as material implication statements. Statistical methods are used to measure the association between the antecedent and the consequent of each postulate. The results show that participants tend to find postulates of update more plausible than postulates of revision.

Acknowledgements

The 7th Workshop on *Formal and Cognitive Reasoning*, which this special issue is dedicated to, attracted researchers from quite different areas of knowledge representation and reasoning, providing the grounds for lots of interesting discussions. We thank all participants of the workshop for their contributions, and we are grateful to Abhaya Nayak for enriching the program by a very inspiring keynote talk. The success of the workshop as well as the selection and careful reviewing of the papers appearing in this issue, would not have been possible without the tremendous support of the program committee members and the additional reviewers; their names are listed on the following page. We would like to give our sincere thanks to all of them. Many thanks also to the organizers of KI-2021 for hosting the workshop at the KI-2021 conference, and to the German Informatics Society. Last but not least, we want to express our gratitude to Dov Gabbay for dedicating an issue of this journal for FCR-2021, and to Jane Spurr for her valuable support in organizing this special issue.

Workshop Organization

Organized by the Special Interest Group on Knowledge Representation and Reasoning and the Special Interest Group on Cognition of the German Informatics Society (GI).

Workshop Organizers and Co-Chairs

Christoph Beierle	FernUniversität in Hagen, Germany
Marco Ragni	Technische Universität Chemnitz, Germany
Frieder Stolzenburg	Hochschule Harz, Germany
Matthias Thimm	FernUniversität in Hagen, Germany

Program Committee

Ringo Baumann	Universität Leipzig, Germany
Christoph Benzmüller	Freie Universität Berlin, Germany
François Bry	LMU München, Germany
Emmanuelle-Anna Dietz	TU Dresden, Germany
Lupita Estefania Gazzo Castaneda	Universität Gießen, Germany
Haythem O. Ismail	German University in Cairo, Egypt
Manfred Kerber	University of Birmingham, UK
Gabriele Kern-Isberner	TU Dortmund, Germany
Steven Kutsch	inovex GmbH, Germany
Sylwia Polberg	Cardiff University, UK
Nico Potyka	Imperial College London, UK
Sebastian Rudolph	TU Dresden, Germany
Ute Schmid	Universität Bamberg, Germany
Claudia Schon	Universität Koblenz-Landau, Germany
Christian Strasser	Ruhr-University Bochum, Germany
Hans Tompits	TU Wien, Austria

Anni-Yasmin Turhan   TU Dresden, Germany
Christoph Wernhard   TU Dresden, Germany
Stefan Woltran   TU Wien, Austria

Additional Reviewers

Clayton Kevin Baker   University of Cape Town, South Africa
Salem Benferhat   Université d'Artois, France
Matti Berthold   Universität Leipzig, Germany
Nir Oren   University of Aberdeen, UK
Ramón Pino Pérez   Université d'Artois, France
Jandson S. Ribeiro   FernUniversität in Hagen, Germany
Kai Sauerwald   FernUniversität in Hagen, Germany
Marco Wilhelm   TU Dortmund, Germany

The Weak Completion Semantics and Counterexamples

Meghna Bhadra[*]
TU Dresden, 01062 Dresden, Germany
meghna.bhadra@tu-dresden.de

Steffen Hölldobler[*]
TU Dresden, 01062 Dresden, Germany and
North Caucasus Federal University, Stavropol, Russian Federation
sh43@posteo.de

Abstract

An experiment has revealed that if the antecedent of a conditional sentence is denied, then most participants conclude that the negation of the consequent holds. However, a significant number of participants answered *nothing follows* if the antecedent of the conditional sentence was *non-necessary*, that is the case when given a conditional *if A then C*, both ($\neg A\ \neg C$) and ($\neg A\ C$) are deemed possible. The Weak Completion Semantics correctly models the answers of the majority, but cannot explain the number of *nothing follows* answers. In this paper we extend the Weak Completion Semantics by counterexamples. The extension allows it to explain the experimental findings.

1 Introduction

Conditional sentences are propositions of the form *if A then C* where A and C are atomic sentences called antecedent and consequent, respectively. Four kinds of conditional inference tasks have been a common area of research by psychologists to date:

1. Affirmation of the antecedent (AA): *if A then C* and A, therefore C.

2. Denial of the antecedent (DA): *if A then C* and $\neg A$, therefore $\neg C$.

---

[*]Authors are mentioned in alphabetical order.

3. Affirmation of the consequent (AC): *if A then C* and *C*, therefore *A*.

4. Denial of the consequent (DC): *if A then C* and $\neg C$, therefore $\neg A$.

In classical, two-valued propositional logic, conditional sentences are taken to mean material implications and bi-conditionals to mean (material) equivalence. The conclusion for the DA and the AC are hence considered to be logical fallacies (invalid) for a conditional sentence whereas they are considered valid for a bi-conditional. However, from our human experiences we know that it is not always the case in real life. When replacing the above abstract conditional sentences which have no everyday context with conditionals which do, the inferences largely depend on the semantics and pragmatics of human communication, culture, and context. In this paper, we therefore discuss how everyday conditional sentences can be categorized into four proposed semantic categories. We also share the results of an experiment reported in [7, 8] and (with particular regard to the DA) demonstrate how such classifications can help model an average human (DA) reasoner.

The Weak Completion Semantics (WCS) is a three-valued, non-monotonic cognitive theory, which can not only adequately model the suppression task by [3] as shown by [9], human syllogistic reasoning as shown by [35], and DC inferences as shown by [8] but also the AA, AC, and the majority $\neg C$ answers of the DA as shown by [7]. While the existing framework of the WCS adequately models the general consensus of the $\neg C$ responses generated in case of the DA inference task, it did not however, seem adequate to model the number of *nothing follows* responses, which is especially significant in case of conditional sentences with *non-necessary antecedents*. Here, *nothing follows* denotes no new inference or specific conclusion can be drawn with regard to the consequent of the conditional sentence.

In order to elaborate on what it really means for a conditional sentence to have a *non-necessary antecedent* and to propose a solution to the aforementioned problem, we begin by considering the following DA inference tasks:

Example 1. *If Maria is drinking alcoholic beverages in a pub, then Maria must be over 19 years of age* and *Maria is not drinking alcoholic beverages in a pub.*

Example 2. *If the plants get water, then they will grow* and *the plants get no water.*

Both of these examples appeared in the aforementioned experiment, and as was the case for every conditional sentence that was included in it, accompanied by a small background story. A curious reader may find the background stories for the above conditionals in the Appendix. In the first example, 28 out of 56 participants answered *Maria is not over 19 years of age*, whereas 25 answered *nothing follows*. In this example the antecedent is non-necessary; it is not considered necessary for a person to drink alcohol in order for her to be older than 19. In fact, there are many people who *do not drink alcoholic beverages* although *they are over 19 years of age*. In the second example, 47 out of 56 participants answered *the plants will not grow* whereas only 8 answered *nothing follows*. In this case, the antecedent is necessary. *Plants do not grow without water.* Table 5 gives a complete account of this experimental data.

Based on this observation, we propose an extension which allows the WCS to account for the *nothing follows* answers. In Example 1, the existing framework of the WCS creates a model where given that *Maria is not drinking alcoholic beverages*, it can be concluded that *Maria is not older than 19 years of age*. With the proposed extension, however, a counterexample that *Maria is not drinking alcoholic beverages* and yet *Maria is older than 19 years of age* can be constructed. This leads to an alternative model, which when compared to the former model and reasoned sceptically, leads to the conclusion that it is unknown whether *Maria is older than 19 years of age*. In Example 2, the WCS creates a model where given that *the plants do not get water*, it can be concluded that *they will not grow*. But in this case, a counterexample does not readily exist.

The paper is organized as follows. Section 2 contains a discussion of some related work; the list, however, is not an exhaustive one. In Section 3 we formally introduce the WCS. A classification of conditional sentences is given in Section 4 where we discuss how conditionals may be classified as obligational or factual, and their antecedents as necessary or non-necessary. We discuss how pragmatics, culture and other such factors may affect how different individuals comprehend the same conditional sentence, and the different possibilities that arise from these comprehensions (in a spirit similar to [22]). Furthermore, we present how the WCS can handle the aforementioned classifications of conditionals. The experiment is described in Section 5. We extend the WCS framework with the search for counterexamples in Section 6. A motivation for modelling the DA inference task is given in Section 7 where we also revisit how the WCS currently models the AC and the DC inference tasks. How the WCS can model the DA is presented in Section 8. This is followed by Section 9 which contains a brief discussion about the predictions of the Mental Model Theory (MMT) and the WCS with regard to the DA, AC and DC. Finally, in Section 10 we conclude and outline possibilities for future research.

2 Related Work

The WCS framework, as the name suggests, utilizes the weak completion of logic programs and in particular their least models in order to model human reasoning scenarios. How logic programs may be weakly completed will be discussed in Section 3. The weak completion of a program denoted by $wc(\mathcal{P})$ differs from a program's completion, $comp(\mathcal{P})$, which involves the following transformation. Given a grounded logic program, $g(\mathcal{P})$,

1. All clauses with the same head such as $A \leftarrow body_1$, $A \leftarrow body_2$, ... are replaced by $A \leftrightarrow body_1 \vee body_2 \vee \ldots$ in $comp(\mathcal{P})$.

2. If an atom B is undefined in $g\mathcal{P}$, i.e. there is no clause such as $B \leftarrow body$, then $comp(\mathcal{P})$ contains $B \leftrightarrow \bot$.

For the moment the said difference between a program completion and weak completion can be illustrated using a simple example. Let $\mathcal{P}'$:

$$\{a \leftarrow \top,\ a \leftarrow b,\ c \leftarrow \bot\}.$$

$wc(\mathcal{P}') = \{a \leftrightarrow \top \vee b,\ c \leftrightarrow \bot\}$, whereas $comp(\mathcal{P}') = \{a \leftrightarrow \top \vee b,\ c \leftrightarrow \bot,\ b \leftrightarrow \bot\}$.

In [11] Melvin Fitting showed that the completion of a logic program admits a least model under the three-valued Fitting logic, the semantics of which is showed in Table 1. This least model can be computed as the least fixed point of the so-called Fitting operator as follows. Given a program $\mathcal{P}$ and an interpretation $I = \langle I^\top, I^\bot \rangle$,[1] the Fitting operator computes positive immediate consequences in line with [1], by mapping an atom A in the grounded program, $g(\mathcal{P})$, to true if there exists a clause of the form $A \leftarrow body$ in $g(\mathcal{P})$ such that $body$ is mapped to true under I. Additionally, it maps a ground atom B to false if $for\ all$ clauses of the form $B \leftarrow body$ in $g(\mathcal{P})$, $body$ has been mapped to false under I. This implies that even in cases when $g(\mathcal{P})$ does not contain a clause of the form $B \leftarrow body$, that is B is undefined in $g(\mathcal{P})$, it will be mapped to false by the Fitting operator.

The Fitting operator was modified by Keith Stenning and Michiel van Lambalgen in [39]. The definition of this modified operator, $\Phi_\mathcal{P}$, for a logic program $\mathcal{P}$, will be presented in Section 3. The difference between the two operators lies in the fact that in case of the latter an atom B in $g(\mathcal{P})$ is mapped to false if and only if *there exists* a rule of the form $B \leftarrow body$, and for all rules of such form we find that $body$ has been mapped to false under the interpretation I. In [39] Stenning and van Lambalgen share some important results of the modified operator, such as its monotonicity and reaching a least fixed point when iterating the operator from the

---

[1]Interpretations are mappings from the set of formulas into the set $\{\top, \mathsf{U}, \bot\}$ where each truth constant denotes true, unknown and false, respectively, such that the truth constants are mapped onto themselves. Interpretations are represented by the two sets $I^\top$ and $I^\bot$ consisting of the set of all ground atoms which are mapped to true and false, respectively. The ground atoms not appearing in the tuple are mapped to unknown.

empty interpretation $\langle \emptyset, \emptyset \rangle$. They also state that the least fixed point of the modified operator can be shown to be a minimal model of a given program $\mathcal{P}$ under the Fitting semantics shown in Table 1. However, the example that they use to illustrate the point, namely a program $\mathcal{P} = \{q \leftarrow p\}$, actually serves as a counter-example to the claim. The reader may observe this when they consider the interpretation $I = \langle \emptyset, \emptyset \rangle$ which is the least fixed point of $\Phi_\mathcal{P}$ but not a minimal model for $\mathcal{P}$. This is because under the Fitting semantics, I maps the clause $q \leftarrow p$ to unknown. The minimal models of $\mathcal{P}$ are instead $\langle \{q\}, \emptyset \rangle$ and $\langle \emptyset, \{p\} \rangle$. Hence, a proposal to solve this problem using Łukasiewicz logic was made in [17]. In particular as shown in [16, 28], the least fixed point of the $\Phi_\mathcal{P}$ operator is equal to the least model of the weak completion of a given program $\mathcal{P}$ under Łukasiewicz logic.

The inference tasks discussed in this paper, i.e. the AA, DA, AC and DC have been modelled based on an experiment conducted with 56 participants, not formally trained in logic; the details of which may be found in the Appendix. The insights from the discussion further helps us understand how theories such as the well-known Mental Model Theory (MMT) and the novel WCS compare to one another with respect to this area of human reasoning. The idea of *"mental models"* is attributed to psychologist Kenneth Craik's work called "The Nature of Explanation" [6], in which he uses the notions of *physical internal models*, probably akin to symbols used in analog devices during that era, in order to discuss cognitive functions in a human brain. He describes the so-called *physical internal models* as *"My hypothesis then is that thought models, or parallels, reality—that its essential feature is not 'the mind', 'the self, 'sense data', nor propositions but symbolism, and that this symbolism is largely of the same kind as that which is familiar to us in mechanical devices which aid thought and calculation"*. His arguments about the human brain's decision-making relying on internal models of the world, led to the inception of the Mental Model Theory [25]. In his paper [24], Philip Johnson-Laird discusses

F	$\neg F$
$\top$	$\bot$
$\bot$	$\top$
U	U

$\wedge$	$\top$	U	$\bot$
$\top$	$\top$	U	$\bot$
U	U	U	$\bot$
$\bot$	$\bot$	$\bot$	$\bot$

$\vee$	$\top$	U	$\bot$
$\top$	$\top$	$\top$	$\top$
U	$\top$	U	U
$\bot$	$\top$	U	$\bot$

$\leftarrow$	$\top$	U	$\bot$
$\top$	$\top$	$\top$	$\top$
U	U	U	$\top$
$\bot$	$\bot$	U	$\top$

$\leftrightarrow$	$\top$	U	$\bot$
$\top$	$\top$	$\bot$	$\bot$
U	$\bot$	$\top$	$\bot$
$\bot$	$\bot$	$\bot$	$\top$

Table 1: The truth tables for Fitting logic. One should observe that $\mathsf{U} \leftarrow \mathsf{U} = \mathsf{U}$ and $\mathsf{U} \leftrightarrow \mathsf{U} = \top$ as shown in the grey cells. Although this logic has already been considered in [30], it has received much attention in the logic programming community after the publication of Fitting's paper [11]. Therefore, we refer to the logic presented herein as *Fitting logic*.

the notion of mental models based on how humans comprehend descriptions of the world, based on their general knowledge, life experiences etc. On a similar note, how mental models of the world are constructed based on what the human vision perceives in front of it has been discussed by David Marr in [34]. Broadly speaking, the MMT is an informal, cognitive theory based on the central idea that much of human reasoning depends on mental models that the brain constructs based on the perception or a description of the real world. Till date, it has discussed various areas of human reasoning in terms of mental models; [26], [19], [24], [20], [23] to name a few. In keeping with the theme of the current paper, some of the predictions of the MMT with regard to the inference tasks will be discussed in Section 9.

Aside from the approaches discussed in this paper, there exist various others that tackle the problems of conditional reasoning, such as [10]. In this paper, the authors substitute the approach of classical logic with one where they instead formalize *inference patterns* which can be deciphered from how reasoners seem to apply or refrain from apparent rules in their responses to the AA, DA, AC and DC inference tasks. Using the formal model of *plausibility relations* based on preferential models [33] and Ordinal Conditional Functions [36, 37], and the constraints imposed by the aforementioned inference patterns on these plausibility relations, they evaluate the rationality of the inference patterns and ultimately that of the individual inferences of the reasoners.

3 The Weak Completion Semantics

We assume the reader to be familiar with logic and logic programming as presented in e.g. [12] and [31]. Let $\top$, $\bot$, and U be truth constants denoting *true, false,* and *unknown*, respectively. A *(logic) program* is a finite set of clauses of the form $B \leftarrow body$, where B is an atom and *body* is $\top$, or $\bot$, or a finite, non-empty conjunction of literals. Clauses of the form $B \leftarrow \top$, $B \leftarrow \bot$, and $B \leftarrow L_1, \ldots, L_n$ are called *facts, assumptions,* and *rules,* respectively, where L_i, $1 \leq i \leq n$, are literals. We restrict our attention to propositional programs although the WCS extends to first-order programs as well [15].

Throughout this paper, $\mathcal{P}$ will denote a program. An atom B is *defined* in $\mathcal{P}$ if and only if $\mathcal{P}$ contains a clause of the form $B \leftarrow body$. As an example consider the program

$$\mathcal{P}_0 = \{l \leftarrow e \wedge \neg ab_e, \; ab_e \leftarrow \bot\},$$

where l, e, and ab are atoms. l and ab_e are defined, whereas e is undefined. ab is an abnormality predicate which is assumed to be false. In the WCS, this program represents the conditional sentence *if A then C*. In their everyday lives humans are often required to reason in situations where the information of all factors affecting the situation might not be complete. They still reason, unless new information which needs consideration comes to light. The abnormality predicate in the program serves the purpose of this (default) assumption, as was suggested in [38].

Now, consider the following transformation:

1. For all defined atoms B occurring in $\mathcal{P}$, replace all clauses of the form $B \leftarrow body_1$, $B \leftarrow body_2$, ... by $B \leftarrow body_1 \vee body_2 \vee \ldots$.

2. Replace all occurrences of $\leftarrow$ by $\leftrightarrow$.

The resulting set of equivalences is called the *weak completion* of $\mathcal{P}$, denoted by $wc\mathcal{P}$.[2] It differs from the program completion defined in [5] in that atoms undefined in a program are not mapped to false in its weak completion, but to unknown instead. Weak completion is necessary for the WCS framework to adequately model the suppression task (and other reasoning tasks) as demonstrated in [9]. As an example of a program and its weak completion, let us reconsider $\mathcal{P}_0$, which has the weak completion $wc\mathcal{P}_0$:

$$\{l \leftrightarrow e \wedge \neg ab_e,\ ab_e \leftrightarrow \bot\}.$$

Within the WCS framework, programs and their weak completions are interpreted under the three-valued Łukasiewicz logic [32] (see Table 2). Let $\mathcal{P}$ be a program and I be a three-valued interpretation represented by the pair $\langle I^\top, I^\bot \rangle$, where $I^\top$ and $I^\bot$ are the sets of atoms mapped to true and false by I, respectively, and atoms which are not listed are mapped to unknown. I is a model for P, in symbols $I \models \mathcal{P}$, if and only if I maps each ground instance of each clause in $\mathcal{P}$ to true. Furthermore, given an interpretation I, if $I = \langle I^\top, I^\bot \rangle \models \mathcal{P}$ then $I' = \langle I^\top, \emptyset \rangle \models \mathcal{P}$. Given two interpretations I_1 and I_2, if $\langle I_1^\top, \emptyset \rangle \models \mathcal{P}$ and $\langle I_2^\top, \emptyset \rangle \models \mathcal{P}$, then $\langle I_1^\top \cap I_2^\top, \emptyset \rangle \models \mathcal{P}$. Importantly, the *model intersection property* holds for $\mathcal{P}$, i.e. $\cap \{I \mid I \models \mathcal{P}\} \models \mathcal{P}$; in other words, the intersection of all models for $\mathcal{P}$ is (also) a model for $\mathcal{P}$, where intersection of two models, such as $I = \langle I^\top, I^\bot \rangle$ and $J = \langle J^\top, J^\bot \rangle$, is defined as $\langle I^\top \cap J^\top, I^\bot \cap J^\bot \rangle$. The aforementioned intersection of all models gives us the *least model* of the program. The model intersection property holds for the weak completion of $\mathcal{P}$, $wc\mathcal{P}$, as well. However, one should note that it is not necessarily the case that the least model of $wc\mathcal{P}$ be the least model for $\mathcal{P}$.

As shown in [16], each weakly completed program admits a least model under the Łukasiewicz logic [32]. This model will be denoted by $\mathcal{M}_{wc\mathcal{P}}$. It can be computed as the least fixed point of a semantic operator introduced in [39]. Let $\mathcal{P}$ be a program

---

[2]Whenever we apply a unary operator like wc to an argument like $\mathcal{P}$, we omit parenthesis and write $wc\mathcal{P}$ instead.

and I be a three-valued interpretation represented by the pair $\langle I^\top, I^\bot \rangle$, where $I^\top$ and $I^\bot$ are the sets of atoms mapped to true and false by I, respectively, and atoms which are not listed are mapped to unknown. We define $\Phi_\mathcal{P} I = \langle J^\top, J^\bot \rangle$, where

$$J^\top = \{B \mid \text{there is } B \leftarrow body \in \mathcal{P} \text{ and } I\, body = \top\},$$
$$J^\bot = \{B \mid \text{there is } B \leftarrow body \in \mathcal{P} \text{ and}$$
$$\qquad\qquad \text{for all } B \leftarrow body \in \mathcal{P} \text{ we find } I\, body = \bot\}.$$

As an example of how such a model can be computed, let us consider the program, $\mathcal{P}'_0 = \mathcal{P}_0 \cup \{e \leftarrow \top\}$. Starting with the interpretation $\langle \emptyset, \emptyset \rangle$, we obtain,

$$\Phi_{\mathcal{P}'_0} \langle \emptyset, \emptyset \rangle = \langle \{e\}, \{ab_e\} \rangle,$$

$$\Phi_{\mathcal{P}'_0} \langle \{e\}, \{ab_e\} \rangle = \langle \{e,\ l\}, \{ab_e\} \rangle = \Phi_{\mathcal{P}'_0} \langle \{e,\ l\}, \{ab_e\} \rangle.$$

Here, $\langle \{e,\ l\}, \{ab_e\} \rangle$ is the least fixed point of $\Phi_{\mathcal{P}'_0}$ and it is also the least model of $wc\mathcal{P}'_0$.

Following [27] we consider an *abductive framework* $\langle \mathcal{P}, \mathcal{A}_\mathcal{P}, \mathcal{IC}, \models_{wcs} \rangle$, where $\mathcal{P}$ is a program, $\mathcal{A}_\mathcal{P} = \{B \leftarrow \top \mid B \text{ is undefined in } \mathcal{P}\} \cup \{B \leftarrow \bot \mid B \text{ is undefined in } \mathcal{P}\}$ is the *set of abducibles*. $\mathcal{IC}$ is a finite set of *integrity constraints* which are expressions of the form $\cup \leftarrow L_1, \ldots, L_n$ and $\bot \leftarrow L_1, \ldots, L_n$ where each L_i, $1 \leq i \leq n$, is a literal. And $\mathcal{M}_{wc\mathcal{P}} \models_{wcs} F$ if and only if $\mathcal{M}_{wc\mathcal{P}}$ maps the formula F to true. Let $\mathcal{O}$ be an *observation*, i.e. a finite set of literals each of which does not follow from $\mathcal{M}_{wc\mathcal{P}}$. We apply abduction to explain $\mathcal{O}$, where $\mathcal{O}$ is called *explainable* in the abductive framework $\langle \mathcal{P}, \mathcal{A}_\mathcal{P}, \mathcal{IC}, \models_{wcs} \rangle$ if and only if there exists a non-empty $\mathcal{X} \subseteq \mathcal{A}_\mathcal{P}$ called an *explanation* such that $\mathcal{M}_{wc(\mathcal{P} \cup \mathcal{X})} \models_{wcs} L$ for all $L \in \mathcal{O}$ and $\mathcal{M}_{wc(\mathcal{P} \cup \mathcal{X})}$ satisfies $\mathcal{IC}$. We have assumed that explanations are non-empty as otherwise the observation already follows from the weak completion of the program. Formula F *follows credulously* from $\mathcal{P}$ and $\mathcal{O}$ if and only if there exists an explanation $\mathcal{X}$ for $\mathcal{O}$ such that $\mathcal{M}_{wc(\mathcal{P} \cup \mathcal{X})} \models_{wcs} F$. F *follows sceptically* from $\mathcal{P}$ and $\mathcal{O}$, if and only if $\mathcal{O}$ can be explained and for all explanations $\mathcal{X}$ for $\mathcal{O}$ we find $\mathcal{M}_{wc(\mathcal{P} \cup \mathcal{X})} \models_{wcs} F$. The latter is an application of the so-called *Gricean implicature* [13]: humans normally

F	$\neg F$
$\top$	$\bot$
$\bot$	$\top$
U	U

$\wedge$	$\top$	U	$\bot$
$\top$	$\top$	U	$\bot$
U	U	U	$\bot$
$\bot$	$\bot$	$\bot$	$\bot$

$\vee$	$\top$	U	$\bot$
$\top$	$\top$	$\top$	$\top$
U	$\top$	U	U
$\bot$	$\top$	U	$\bot$

$\leftarrow$	$\top$	U	$\bot$
$\top$	$\top$	$\top$	$\top$
U	U	$\top$	$\top$
$\bot$	$\bot$	U	$\top$

$\leftrightarrow$	$\top$	U	$\bot$
$\top$	$\top$	U	$\bot$
U	U	$\top$	U
$\bot$	$\bot$	U	$\top$

Table 2: The truth tables for Lukasiewicz logic. One should observe that $U \leftarrow U = U \leftrightarrow U = \top$ as shown in the grey cells.

do not quantify over things which do not exist. Meaning, (unlike classical logic) *all* explanations for an observation $\mathcal{O}$ may only be taken into account to sceptically decide on a formula F, when $\mathcal{O}$ is explainable and these so-called explanations exist in the first place. If a formula F does not follow sceptically from $\mathcal{P}$ and $\mathcal{O}$, we conclude *nothing follows*. Furthermore, one should also observe that if an observation $\mathcal{O}$ cannot be explained, then *nothing follows* credulously as well as sceptically. In all examples discussed in this paper the set of integrity constraints is empty. Integrity constraints are not very relevant to the goal of this paper but they are in other applications of the WCS like human disjunctive reasoning [14].

Given premises and general knowledge encoded as a logic program, and observations encoded as a finite set of ground literals, *reasoning in the WCS* is currently modelled in five steps:

1. Reasoning towards a logic program $\mathcal{P}$ following [39].

2. Weakly completing the program, which leads to $wc\mathcal{P}$.

3. Computing the least model $\mathcal{M}_{wc\mathcal{P}}$ of the weak completion of $\mathcal{P}$, $wc\mathcal{P}$, under the three-valued Łukasiewicz logic.

4. Reasoning with respect to $\mathcal{M}_{wc\mathcal{P}}$.

5. If observations cannot be explained, then applying sceptical abduction using the specified set of abducibles.

In the following sections we will explain how these five steps work in the case of the DA reasoning tasks considered in this paper. More examples can be found, for example, in [9] or [35] or [14].

4 A Classification of Conditional Sentences

4.1 Obligational versus Factual Conditionals

Following [4], we call a conditional sentence an *obligational conditional* if the truth of the consequent appears to be obligatory given that its antecedent is true. For each obligational conditional there are two initial possibilities humans think about. The first possibility is the conjunction of the antecedent and the consequent, which is permitted. The second possibility is the conjunction of the antecedent and the negation of the consequent, which is forbidden. Exceptions are possible but unlikely. This can be exemplified by Example 1. In many countries the law demands that a person may only drink alcohol publicly when they are above a certain age group (for example, 19 years). This implies that *Maria is drinking alcoholic beverages in a pub* and *she is older than 19 years* is a permitted possibility, whereas *Maria is drinking alcoholic beverages in a pub* and *she is not older than 19 years* is a forbidden one. Hence, *if Maria is drinking alcoholic beverages in a pub, then Maria must be over 19 years of age* is an obligational conditional. In Example 2, *plants getting water* and *plants are growing* is a permitted possibility. But *plants getting water* and *plants are not growing* is also possible; there are many other factors, for example, overwatering, lack of light, pest infestation, etc. which may hinder their growth. Hence, *if the plants get water, then they will grow*, is not an obligational conditional.

Obligational conditionals may have different sources. They may be based on legal laws like Example 1 and are often called deontic conditionals, in which case words like *must, should* or *ought* may be explicitly used in the conditional sentence. Their usage however, does not seem mandatory in everyday communication and is skipped on many occasions. Knowledge or awareness that the consequent is obligatory given the antecedent suffices in these cases, and yields the same responses as when explicitly denoting the obligation. Obligational conditionals may also express moral or social obligations like *if somebody's parents are elderly, then he/she should look after them* [4]. Other obligational conditionals are based on causal or physical laws which hold on our planet like, *if an object is not supported, then it will fall to the ground*. In each case the conjunction of the antecedent and the consequent is permitted, whereas the conjunction of the antecedent and the negation of the consequent is forbidden.

On the other end of the spectrum, if the consequent of a conditional sentence is not obligatory given the antecedent, then it is called a *factual conditional*. In particular, the truth of the antecedent is *inconsequential* to that of the consequent; that is (even) if the antecedent is true, the consequent may or may not be true. This has already been exemplified using Example 2; the conditional *if the plants get water, then they will grow* is a factual one. As another example consider the conditional sentence *if Maria is over 19 years, then she might drink alcoholic beverages in a pub*. This sentence is a factual one, because given the atomic proposition *Maria is over 19 years* is true, one can imagine two permitted possibilities, one where *Maria drinks alcohol beverages* and another where *Maria does not drink alcoholic beverages in a pub*.

4.2 Necessary versus Non-Necessary Antecedents

As discussed in the previous sub-section, the obligational or factual nature of a conditional sentence indicates if the consequent is obligatory or simply possible, provided the antecedent is satisfied. The question that may naturally arise at this point is, what happens when the antecedent of a conditional sentence is not satisfied? To that end, we now discuss the classifications of antecedents of conditional sentences. The antecedent A of a conditional sentence *if A then C* is said to be *necessary* with respect to the consequent C, if and only if C cannot be true unless A is true. This implies that if A does not hold, C cannot either. In Example 2, *plants get water* is a necessary antecedent for *plants will grow*. If a plant is not watered at all, it will very likely die.

The above does not imply however, that the antecedent need always be a precondition for the consequent, per se. The antecedent A of a conditional sentence *if A then C* is said to be *non-necessary* with respect to the consequent C, if C can be true irrespective of the truth or falsity of A. In particular this implies, if A does not hold, C may or may not hold. In Example 1, the falsity of *drinking alcoholic beverages in a pub* is inconsequential to the truth of the consequent *older than 19 years*. There are plenty of adults (over 19 years) who do not drink alcohol. The antecedent of the conditional sentence *if Maria is drinking alcoholic beverages in a pub, then Maria must be over 19 years of age*, in Example 1 is therefore called non-necessary.

4.3 Pragmatics

Generally, humans may recognize conditional sentences as obligational or factual and antecedents as necessary or non-necessary. This leads to an informal and pragmatic classification of four kinds: obligational conditional with necessary antecedent (ON) or non-necessary antecedent (ONN) and factual conditional with necessary antecedent (FN) or non-necessary antecedent (FNN). For an abstract conditional *if A then C*, *without* an everyday context, the classification of the conditional into any of the aforementioned kinds would be straightforward and as discussed in Subsections 4.1 and 4.2, since they are independent of context. The classification of everyday conditionals (those *with* an everyday context), however, often depend on pragmatics: the context, the background knowledge and experience of a person. For example, the conditional sentence *if it is cloudy, then it is raining* discussed in [29] may be classified as an obligational conditional with necessary antecedent by people living in Java, whereas it may be classified as a factual conditional by people living in Central Europe. In another example [22], the authors conducted an experiment, where they categorized the proposition *if it's heated, then this butter will melt* as a bi-conditional. In particular they considered *if butter is not heated, it will not melt*. This corresponds to a necessary antecedent in our setting. While some of their participants also gave it the same classification, many considered it possible that even *if butter is not heated (intentionally), it may still melt*. This implies that they considered the antecedent to be non-necessary.

4.4 Possibilities Arising from the Classifications of Conditional Sentences

In their paper [22], dedicated to the MMT and the meanings of conditionals, the authors discuss the notion of *sets of possibilities* arising from conditional sentences. Simply put, given a conditional *if A then C*, how humans comprehend or understand it depends on the following questions: what are the possibilities of C when A is satisfied, and when A is not satisfied? This notion of possibilities was harnessed for a detailed comparison between the MMT and the WCS in [2] and is not in the scope of the present discussion. For our current purposes, we limit our attention to characterizing the discussions in Subsections 4.1 and 4.2 using Table 3. It illustrates the meanings of the classifications ON, ONN, FN and FNN in terms of possibilities using the literals A, $\neg A$, C, and $\neg C$ in lines with the MMT. For the moment, we leave out the cases where any of the aforementioned literals may be unknown. The sets of possibilities that an individual may use to characterize a conditional sentence may differ from another individual, depending upon factors like pragmatics, culture, context etc. as was discussed in Subsection 4.3.

4.5 Handling Classifications in the WCS

In the WCS framework, the classification of conditional sentences can be taken into account by extending the definition of the set of abducibles to,

$$\mathcal{A}_{\mathcal{P}}^{e} = \mathcal{A}_{\mathcal{P}} \cup \mathcal{A}_{\mathcal{P}}^{nn} \cup \mathcal{A}_{\mathcal{P}}^{f},$$

	if A then C		
ON	ONN	FN	FNN
A C	A C	A C	A C
$\neg A$ $\neg C$	$\neg A$ $\neg C$	A $\neg C$	A $\neg C$
	$\neg A$ C	$\neg A$ $\neg C$	$\neg A$ $\neg C$
			$\neg A$ C

Table 3: All possibilities of antecedent A and consequent C for each classification of everyday conditional sentences, *if A then C*.

where $\mathcal{A}_\mathcal{P}$ is as defined earlier in Section 3 and,

$$\mathcal{A}_\mathcal{P}^{nn} = \{C \leftarrow \top \mid C \text{ is the head of a rule occurring in } \mathcal{P} \text{ representing a} \\ \text{conditional sentence with non-necessary antecedent}\},$$

$$\mathcal{A}_\mathcal{P}^{f} = \{ab \leftarrow \top \mid ab \text{ occurs in the body of a rule occurring in } \mathcal{P} \\ \text{representing a factual conditional}\}.$$

The set $\mathcal{A}_\mathcal{P}^{nn}$ contains facts for the consequents of conditional sentences with non-necessary antecedents. As was mentioned earlier, if an antecedent of a conditional sentence is non-necessary then the truth of the consequent does not depend on the truth of the antecedent. The abducible $C \leftarrow \top$ therefore implies that there may be other unknown reasons for establishing the consequent of the conditional sentence.

The set $\mathcal{A}_\mathcal{P}^{f}$ contains facts for the abnormality predicates occurring in the bodies of the (logic program) representation of factual conditionals. Owing to the factual nature of a conditional sentence, the antecedent of the conditional may be true, however its consequent may not hold, due to various reasons which we might broadly call abnormalities. As mentioned earlier, considerations of other plausible factors at play might override our default assumption that these abnormalities are false. Once we weakly complete our program, the abducible $ab \leftarrow \top$ shall cause the abnormality predicate to become true and its negation to become false. Hence, the body of the clause containing its negation will be false, causing the consequent to be false in turn. This technique is used in [9] to represent an enabling relation and model, for example, the suppression effect during the AA inference in the suppression task [3]. The original task was as follows. Given, *if she has an essay to write then she will study late in the library, if the library is open then she will study late in the library* and *she has an essay to write*. Only 38% of the participants in the original experiment had responded *she will study late in the library*. Although the percentage of participants who responded otherwise was not revealed in the paper, it is plausible that many considered that *a library not being open* prevents a person from *studying in it*. This can be modelled using the abnormality predicate. Table 4 illustrates how the set of abducibles can be extended for each classification.

5 An Experiment

In [7, 8] an experiment concerning conditional reasoning is described, where 56 logically naive participants were tested on an online website (Prolific, `prolific.co`). The participants were restricted to Central Europe and Great Britain as they were assumed to have similar background knowledge about weather etc. It was also assumed that the participants had not received any education in logic beyond high school training. During the experiment, the participants were presented with a story followed by a first assertion (a conditional premise), and a second assertion (a possibly negated atomic premise). Finally for each problem they had to answer the question "What follows?". Both parts were presented simultaneously. The participants responded by clicking one of the answer options. They could take as much time as they needed and acted as their own controls.

The participants carried out 48 problems consisting of the 12 conditionals listed in the Appendix and solved all four inference types (AA, DA, AC, DC). They could select one of three responses: *nothing follows*, the fact that had not been presented in the second premise, and the negation of this fact. For example in the case of the DA, the first assertion was of the form *if A then C*, the second assertion was $\neg A$, and they could answer C, $\neg C$, or *nothing follows*. It should also be mentioned that the classification of the conditional sentences into the four aforementioned kinds, such as obligational conditional with necessary antecedent, factual conditional with non-necessary antecedent etc., was done by the authors of the experiment and not revealed to the participants.

We can exemplify all that has been said above with the following short scenario taken from the experiment: *Peter has a lawn in front of his house. He is keen to make sure that the grass on the lawn does not dry out, so whenever it has been dry for multiple days, he turns on the sprinkler to water the lawn.* Along with this context the conditional sentence *if it rains, then the lawn is wet* and the negated atomic proposition *it does not rain* were provided. The participants were given three choices of answers: *the lawn is wet, the lawn is not wet*, and *nothing follows*.

As mentioned earlier, the WCS could well explain the findings of the experiment in the cases AA, AC, and DC (see [7, 8]), but failed to explain the findings in the case of DA. The data is shown in Table 5, where the total number of selected responses as well as the median response time in milliseconds for $\neg C$ (*Mdn $\neg C$*) and *nothing follows* (*Mdn nf*) responses are listed. Everyday contexts for the DA inference task elicited a high response rate of about 78% (525 out of 672) for $\neg C$, but in case of

$C \leftarrow A \wedge \neg ab$	non-necessary A	necessary A
factual conditional	$ab \leftarrow \top$, $C \leftarrow \top$	$ab \leftarrow \top$
obligational conditional	$C \leftarrow \top$	

Table 4: The additional facts in the set of abducibles for a rule of the form $C \leftarrow A \wedge \neg ab$ representing a conditional sentence *if A then C*.

nothing follows the rate varied from 8% (14 out of 168) up to 33% (56 out of 168). The number of participants answering C seems irrelevant. Until the present, the WCS could predict the $\neg C$ answered by the majority of the participants, but it could not yet model the significant number of *nothing follows* responses. We now propose a solution to the latter. Before we elaborate further, one might first observe that as per the data *nothing follows* was answered much more often in case of conditional sentences with non-necessary antecedents than in the case of conditional sentences with necessary ones (30% vs. 8%, Wilcoxon signed rank, $W = 0$, $p < .001$). More importantly, the reader may observe that when the classification of the antecedents changed from *necessary* to *non-necessary* the number of $\neg C$ responses decreased to 225 and *nothing follows* increased to (a significant) 101. **The goal of this paper is to extend the WCS in order to model this observed phenomenon.**

6 Extending the WCS to Search for Counterexamples

As shown in Table 5 the majority of the participants always answered $\neg C$ when given the premises *if A then C* and $\neg A$. The classification of conditional sentences seems irrelevant at this point. However, this general consensus is sometimes only barely met. Indeed, some humans seem to be responding *nothing follows* during the DA task. Upon a closer look at Table 5, the reader may observe that the number of *nothing follows* responses increases when the classification of the antecedent of the conditional changes from necessary to non-necessary. This is because unlike a necessary antecedent, a non-necessary one makes room for counterexamples where even if the antecedent does not hold, the consequent might still hold (that is $\neg A$ and C is possible). This observation hints at two reasoning patterns. The first, whom we may term as a *general reasoner*, i.e. one who responds $\neg C$ to any DA inference task and does not deliberate upon it, and the second, the reasoner who does. The latter, whom we may also call the *careful reasoner* searches for counterexamples before drawing a definite conclusion such as $\neg C$, unlike the former who does not. The said counterexamples in the DA task are possible when an individual deems the antecedent to be non-necessary.

The aforementioned difference between the two kinds of reasoning patterns is unfortunately not very noticeable if the so-called careful reasoner has deliberated upon the problem but considered the antecedent to be necessary. In what follows, we attempt to clarify this statement for the reader while also illustrating an approach to model the general consensus of $\neg C$. For example, consider Example 2 ((8) in Table 5), for which 47 participants responded $\neg C$ while only 8 responded *nothing follows.* Assuming it is known that *the plants do not get water* we obtain the program

$$\mathcal{P}_1 = \{g \leftarrow w \wedge \neg ab_1,\ ab_1 \leftarrow \bot,\ w \leftarrow \bot\},$$

where g and w denote that *the plants will grow* and *the plants get water*, respectively, and ab_1 is an abnormality predicate which is implicitly assumed to be false. Weakly

Conditional/Classification	C	pct.	¬C	pct.	nf	pct.	Sum	Mdn ¬C	Mdn nf
(1)	0		45		11		56	2863	4901
(2)	2		54		0		56	3367	na
(3)	2		51		3		56	3647	10477
ON	4	2%	150	89%	14	8%	168	3292	7689
(4)	1		40		15		56	3722	7189
(5)	3		28		25		56	5735	7814
(6)	4		36		16		56	3602	6240
ONN	8	5%	104	62%	56	33%	168	4353	7081
(7)	2		51		3		56	3928	7273
(8)	1		47		8		56	3296	5728
(9)	1		52		3		56	3549	8735
FN	4	2%	150	89%	14	8%	168	3591	7245
(10)	1		39		16		56	3725	6874
(11)	0		41		15		56	3374	5887
(12)	1		41		14		56	3205	7002
FNN	2	1%	121	72%	45	27%	168	3435	6588
Obligational Conditional (O)	12	4%	254	76%	70	21%	336	3823	7385
Factual Conditional (F)	6	2%	271	81%	59	18%	336	3513	6917
Necessary Antecedent (N)	8	2%	300	89%	28	8%	336	3442	7467
Non-Necessary Antecedent (NN)	10	3%	225	67%	101	30%	336	3894	6835
Total	18	3%	525	78%	129	19%	672	3668	7151

Table 5: The results for DA inferences given a conditional sentence *if A then C* and a negated atomic sentence $\neg A$. The grey lines show the numbers for the examples discussed in the introduction. If the antecedent is non-necessary, then *nothing follows (nf)* is answered significantly often (gray cells at the bottom). ON: obligational conditional with necessary antecedent, ONN: obligational conditional with non-necessary antecedent, FN: factual conditional with necessary antecedent, and FNN: factual conditional with non-necessary antecedent. All percentages (*pct.*) have been rounded off to the nearest natural number for the convenience of the reader.

completing $\mathcal{P}_1$ we obtain:

$$\{g \leftrightarrow w \land \neg ab_1,\ ab_1 \leftrightarrow \bot,\ w \leftrightarrow \bot\},$$

whose least model is

$$\mathcal{M}_{wc\mathcal{P}_1} = \langle \emptyset, \{g, ab_1, w\}\rangle,$$

where nothing is true, and g, ab_1, and w are all false. Because the antecedent, *the plants get water* (w), is generally considered to be *necessary* for the consequent, *plants will grow* (g), the falsity of w allows us to falsify g. Hence, we conclude that *the plants will not grow*. This is the general consensus for this particular example. Please note that the authors of the experiment classified this conditional as FN, that is, the antecedent as necessary. Although it is difficult to ascertain how humans comprehend conditionals without inquiring of them, it is plausible that some careful reasoners who deliberate upon this DA task may not find a counterexample where the plants receive no water but they still grow, that is the possibility,

$$\neg A \quad C.$$

In other words, these reasoners comprehend the antecedent to be necessary for the consequent and so conclude $\neg C$, much like the general reasoner. In such a case, there seems to be no apparent way to distinguish between the general and the careful reasoner.

Now we discuss the case of the non-necessary antecedent in the DA task where the difference between the reasoning patterns is more pronounced, and how the WCS may model the two. Let us reconsider Example 1 ((5) in Table 5) where 28 out of 56 participants answered $\neg C$, whereas 25 participants answered *nothing follows*. Interestingly, the latter took more time in their response compared to the former. For these 25 participants the antecedent of the conditional may plausibly have been non-necessary for the consequent. That is upon deliberation, they may have been able to construct a counterexample to the putative conclusion $\neg C$, where *Maria is not drinking alcoholic beverages in a pub* but she may nevertheless be *over 19 years of age*. Because Maria may simply abstain from alcohol.

Overall, the data in Table 5 suggests that the difference between the reasoning patterns during a DA inference task becomes more apparent when the antecedent of the conditional is taken to be non-necessary as it leads to the possibility of counterexamples. In this paper we hence propose to model DA inferences by extending the WCS with the addition of a sixth step to the procedure presented in Section 3:

1. Reasoning towards a logic program $\mathcal{P}$ following [39].

2. Weakly completing the program, which leads to $wc\mathcal{P}$.

3. Computing the least model $\mathcal{M}_{wc\mathcal{P}}$ of the weak completion of $\mathcal{P}$, $wc\mathcal{P}$, under the three-valued Łukasiewicz logic.

4. Reasoning with respect to $\mathcal{M}_{wc\mathcal{P}}$.

5. If observations cannot be explained, then applying sceptical abduction using the specified set of abducibles.

6. **Search for counterexamples.**

The sixth step corresponds to the validation step in the Mental Model Theory [21] in that alternative models falsifying a putative conclusion are searched for. Particularly in the case of the DA task, $\neg C$ may be considered as the putative conclusion generated due to steps 1 to 5. In the sixth step, using the extended set of abducibles $\mathcal{A}_{\mathcal{P}}^{e}$ illustrated in Sub-section 4.5, the extended procedure searches for models where $\neg A$ is true, but $\neg C$ is not. If such models are found, then sceptical reasoning with respect to all constructed models is applied. This will be illustrated and discussed in more detail in the next section.

7 Motivation for the DA Inference Task

In order to discuss how the WCS along with its extension can model the general consensus of $\neg C$ in the DA task and also explain the significant number of *nothing follows* answers in case of non-necessary antecedents, we return to Example 1 and assume that *Maria is not drinking alcoholic beverages in a pub*. In the WCS this is formalized by

$$\mathcal{P}_2 = \{o \leftarrow a \wedge \neg ab_2,\ ab_2 \leftarrow \bot,\ a \leftarrow \bot\},$$

where o and a denote that *Maria is over 19 years old* and *she is drinking alcoholic beverages*, respectively, and ab_2 is an abnormality predicate which is initially assumed to be false. As the weak completion of $\mathcal{P}_2$ we obtain

$$\{o \leftrightarrow a \wedge \neg ab_2,\ ab_2 \leftrightarrow \bot,\ a \leftrightarrow \bot\},$$

whose least model is

$$\mathcal{M}_{wc\mathcal{P}_2} = \langle \emptyset, \{a, ab_2, o\} \rangle.$$

Here, a, ab_2, and o are all false.

A general reasoner following this approach will draw the conclusion *Maria is not over 19 years old* and stop reasoning at this point. This accounts for the 28 $\neg C$

The Weak Completion Semantics and Counterexamples

responses for this particular conditional in our data. Overall it so appears that these participants treated the conditional sentence as a bi-conditional, hence considering only the possibilities corresponding to ON in Table 3, namely,

$$\begin{array}{cc} a & o \\ \neg a & \neg o. \end{array}$$

Classifying an antecedent as non-necessary however, would also allow the consequent to be true despite the falsity of the former. In other words, recognizing an antecedent as non-necessary, might allow humans to consider two possibilities: *Maria does not drink alcohol in a pub* and *she is younger than 19 years*, and *Maria does not drink alcohol in a pub* but *she is older than 19*. Meaning, these participants did not treat the conditional sentence as a bi-conditional, but instead regarded the possibilities,

$$\begin{array}{cc} a & o \\ \neg a & \neg o \\ \neg a & o \end{array}$$

corresponding to that of ONN in Table 3. That is, careful reasoners not only consider the aforementioned model $\mathcal{M}_{wc\mathcal{P}_2}$, where o is mapped to false, but also search for a counterexample to $\neg o$. Investigating the third possibility listed above may lead them to

$$\langle \{o\}, \{a, ab_2\} \rangle,$$

which is also a model for the program $\mathcal{P}_2$, but not a model for $wc\mathcal{P}_2$. **The question now stands, how can this (and similar) counterexamples in the DA inference task be modelled by the WCS?** We would like to motivate a plausible answer to this by first turning our attention to how the WCS models the AC and the DC inference tasks.

7.1 Modelling the AC Inference Task

To illustrate how the WCS models the AC inference task we reconsider the previously discussed conditional sentence,

if Maria is drinking alcoholic beverages in a pub, then Maria must be over 19 years of age

and the atomic premise,

Maria is over 19 years of age.

Unlike the DA, in this case only the conditional premise is represented as a logic program, that is,

$$\mathcal{P}_3 = \{o \leftarrow a \wedge \neg ab_2, \ ab_2 \leftarrow \bot\},$$

where o and a denote that *Maria is over 19 years old* and *she is drinking alcoholic beverages*, respectively. ab_2 is the abnormality predicate. The atomic premise, o, is not considered as a fact because the program $\mathcal{P}_3$ already contains a definition of o and the addition of the fact $o \leftarrow \top$ would override this definition upon weak completion thus not giving us much information about a. Therefore, o is considered as an observation that needs to be explained. Meaning, we apply abduction. Because a is undefined in $\mathcal{P}_3$ and the conditional sentence is classified as obligational with a *non-necessary* antecedent, we obtain,

$$\mathcal{A}_{\mathcal{P}_3} = \{a \leftarrow \top, \ a \leftarrow \bot\} \quad \text{and} \quad \mathcal{A}^e_{\mathcal{P}_3} = \mathcal{A}_{\mathcal{P}_3} \cup \{o \leftarrow \top\}$$

respectively. The set of integrity constraints is empty. Considering $\mathcal{A}_{\mathcal{P}_3}$, the observation o is explained by the minimal explanation

$$\{a \leftarrow \top\}. \tag{1}$$

Adding this explanation to $\mathcal{P}_3$, weakly completing the extended program, and computing its least model, a general reasoner obtains,

$$\langle\{o, a\}, \{ab_2\}\rangle \tag{2}$$

and concludes that *Maria is drinking alcoholic beverages in a pub*. However, a careful reasoner additionally searching for counterexamples may discover a second explanation to the observation, viz.

$$\{o \leftarrow \top\} \tag{3}$$

by considering $\mathcal{A}^e_{\mathcal{P}_3}$. Here, o being true signifies the possibility that *Maria might still be over 19 irrespective of whether she is drinking alcohol in a pub or not*. Adding such an explanation to $\mathcal{P}_3$, weakly completing the extended program, and computing its least model the careful reasoner obtains,

$$\langle\{o\}, \{ab_2\}\rangle. \tag{4}$$

Comparing the least models (2) where a is true and (4) where a is unknown, and reasoning sceptically, a careful reasoner concludes *nothing follows*. We would like to point out to the reader that the explanations (1) and (3) are independent in that neither is a subset nor a superset of the other.

To summarize, it appears that the general reasoner considers $\mathcal{A}_{\mathcal{P}_3}$ and concludes that a and o hold. On the other hand, the careful reasoner considers $\mathcal{A}^e_{\mathcal{P}_3}$, investigates counterexamples, reasons that although o holds a need not necessarily hold, and finally concludes *nothing follows*. Table 6 in the Appendix illustrates the results for the AC task for all conditional sentences used in the experiment. Overall as an investigation of the table would suggest, like in the DA, it is the (necessary or non-necessary) type of the antecedent of the conditional which plausibly influences the search for counterexamples. This is also supported by the time measured for the conditional in the experiment (see (5) in Table 6), the answer a had a median response time of 4704 ms, whereas the answer *nothing follows* had a median response time of 6044 ms.

7.2 Modelling the DC Inference Task

Here we discuss how the WCS framework currently models the DC inference task. For this purpose, let us consider the conditional sentence,

if Ron scores a goal, then he is happy

and the atomic sentence,

Ron is not happy.

The following program represents the conditional premise

$$\mathcal{P}_4 = \{h \leftarrow g \wedge \neg ab_3,\ ab_3 \leftarrow \bot\},$$

where g and h denote *Ron scores a goal* and *Ron is happy*, respectively, and ab_3 is an abnormality predicate. Here, $\neg h$ is considered as an observation that needs to be explained because the program $\mathcal{P}_4$ already contains a definition for h. Adding $h \leftarrow \bot$ to the program has no effect as it will be overridden once the program is weakly completed. As g is undefined and the conditional sentence is classified as a *factual* conditional with non-necessary antecedent, therefore,

$$\mathcal{A}_{\mathcal{P}_4} = \{g \leftarrow \top,\ g \leftarrow \bot\} \quad \text{and} \quad \mathcal{A}^e_{\mathcal{P}_4} = \mathcal{A}_{\mathcal{P}_4} \cup \{h \leftarrow \top,\ ab_3 \leftarrow \top\},$$

respectively. The set of integrity constraints is empty. Considering $\mathcal{A}_{\mathcal{P}_4}$ the observation $\neg h$ is explained by the minimal explanation

$$\{g \leftarrow \bot\}. \tag{5}$$

Adding this explanation to $\mathcal{P}_4$, weakly completing the extended program, and computing its least model a general reasoner will obtain

$$\langle \emptyset, \{h, g, ab_3\}\rangle \quad (6)$$

and conclude, that *Ron does not score a goal*. This is where most reasoners seem to halt their reasoning. However, there may be some individuals who recognize the conditional sentence as factual, meaning, they recognize that $\neg h$ need not just be caused or explained by $\neg g$. More precisely, a careful reasoner will recognize that h may be false, even if g is not. Analogously, such a reasoner will search for counterexamples to the putative conclusion $\neg g$, which can also explain $\neg h$. Using $\mathcal{A}^e_{\mathcal{P}_4}$,

$$\{ab_3 \leftarrow \top\} \quad (7)$$

can be used as an another minimal explanation for $\neg h$. Here ab_3 being true indicates that *Ron may have other reasons to be unhappy*. Adding this abducible to $\mathcal{P}_4$ and weakly completing the resulting extended program leads to the least model

$$\langle \{ab_3\}, \{h\}\rangle. \quad (8)$$

As g is false in the first model (6) whereas unknown in the second, (8), sceptical abduction is applied which leads to the conclusion, *nothing follows*. The overall point of importance is that, aside from the falsity of g it is also possible to find other reasons which can cause h to be false, and this leads to the consideration of more than one model, which may lead humans to reason sceptically.

To summarize, we stipulate that in the above case the general reasoner considers $\mathcal{A}_{\mathcal{P}_4}$ and concludes that given $\neg h$, $\neg g$ holds, whereas a careful reasoner considers $\mathcal{A}^e_{\mathcal{P}_4}$, finds a counterexample where $\neg h$ holds and $\neg g$ is unknown, and sceptically concludes *nothing follows*. In case of the DC inference task it is the (obligational or factual) type of the conditional which plausibly influences this said search for counterexamples. Comparing the explanations (5) and (7) we find that they are independent of each other. It must also be pointed out that the average time taken by participants to respond $\neg A$, for this particular task (see (12) in Table 7), was 3726 ms and that to respond *nothing follows* was 3813 ms, which are quite comparable.

8 Modelling the $\neg C$ and Nothing Follows Responses in the DA Task

In Subsections 7.1 and 7.2, when modelling the general consensus as well as the *nothing follows* responses in the AC and DC inference tasks, the premise C and $\neg C$

are considered as observations, respectively. The *nothing follows* responses can be accounted for by using the extended set of abducibles and applying sceptical abduction in order to explain the observation. This may lead to models which act as counterexamples to each other. On the other hand, when modelling the general consensus for the DA, the atomic premise was a part of the logic program representation of the premises, as illustrated in the beginning of Section 7.

We now propose that this negated premise be considered as an observation instead, in a fashion similar to the AC and the DC modelling techniques. To illustrate the proposal we reconsider the premises,

> *if Maria is drinking alcoholic beverages in a pub, then Maria must be over 19 years of age*

and,

> *Maria is not over 19 years of age.*

In this revised proposal, the conditional premise is represented by

$$\mathcal{P}_5 = \{o \leftarrow a \wedge \neg ab_2,\ ab_2 \leftarrow \bot\},$$

where the meanings of the atomic predicates are unchanged. Now $\neg a$ is considered as an observation. As was also mentioned earlier, it is the (necessary or non-necessary) type of the antecedent of the conditional which plausibly influences the search for counterexamples in a DA task. Since this particular conditional sentence is classified as obligational with non-necessary antecedent, we obtain,

$$\mathcal{A}_{\mathcal{P}_5} = \{a \leftarrow \top,\ a \leftarrow \bot\} \quad \text{and} \quad \mathcal{A}^e_{\mathcal{P}_5} = \mathcal{A}_{\mathcal{P}_5} \cup \{o \leftarrow \top\}.$$

The set of integrity constraints is empty. Considering $\mathcal{A}_{\mathcal{P}_5}$ the observation is explained by the minimal explanation,

$$\{a \leftarrow \bot\}. \tag{9}$$

The reader may note that adding this explanation to $\mathcal{P}_5$ leads to the program $\mathcal{P}_2$ introduced at the beginning of Section 7. Again, weakly completing $\mathcal{P}_2$ and computing its least model we obtain,

$$\langle \emptyset, \{o, ab_2, a\} \rangle. \tag{10}$$

Hence the conclusion is, that *Maria is not older than 19 years of age*. The data in Table 5 suggests that many reasoners stop reasoning at this point, thereby treating the conditional premise as a bi-conditional. Following this revised technique allows us to model the general consensus or the general reasoner.

Now we turn our attention to the careful reasoner, the one who searches for counterexamples to the aforementioned conclusion and, in particular, considers $\mathcal{A}^e_{\mathcal{P}_5}$. Such a reasoner may discover a second, non-minimal explanation to $\neg a$, viz.

$$\{a \leftarrow \bot, \ o \leftarrow \top\}. \tag{11}$$

This translates to the possibility that *Maria is not drinking alcoholic beverages but she is over 19 years of age*. Adding this explanation to $\mathcal{P}_5$, weakly completing the extended program, now leads us to the least model,

$$\langle \{o\}, \{ab_2, a\} \rangle. \tag{12}$$

Comparing the least models (10) where o is false and (12) where o is true and reasoning sceptically, one concludes *nothing follows*. Comparing the explanations (9) and (11) we find

$$\{a \leftarrow \bot\} \subset \{a \leftarrow \bot, \ o \leftarrow \top\},$$

meaning, the second explanation is a superset of the first.

To summarize, in a manner similar to the AC and DC, the second premise in a DA reasoning task is considered as an observation which needs to be explained. The general reasoner considers $\mathcal{A}_{\mathcal{P}_5}$ and concludes that given $\neg a$, $\neg o$ holds. On the other hand, a careful reasoner considers $\mathcal{A}^e_{\mathcal{P}_5}$, reasons that although $\neg a$ holds, o may nevertheless hold, and sceptically concludes *nothing follows*. This is also supported by the time measured in the experiment as shown in Table 5 (see (5)). The answer $\neg o$ had a median response time of 5735 ms, whereas the answer *nothing follows* had a median response time of 7814 ms.

9 A Brief Discussion about the Predictions of the Mental Model Theory

The scope of the MMT is broad and has been applied to quite a few areas of human reasoning to date. At present we restrict ourselves to a brief discussion of some of the predictions of the MMT, as discussed by Philip Johnson-Laird and Ruth Byrne, in their paper [22] in case of the inference tasks, DA, AC and DC. The MMT suggests that in a DA task, given a conditional sentence *if A then C* and $\neg A$, humans *intuitively* refrain from responding $\neg C$ and favour *nothing follows*. It thus predicts that when reasoners respond with $\neg C$, it is a result of deliberately comprehending the conditional sentence as a bi-conditional, meaning the possibilities,

$$\begin{array}{cc} A & C \\ \neg A & \neg C. \end{array}$$

However, in the case that a reasoner comprehends a conditional sentence as a conditional, meaning the possibilities,

$$\begin{array}{cc} A & C \\ \neg A & \neg C \\ \neg A & C \end{array}$$

the reasoner refrains from the response of $\neg C$ upon deliberation. Now, if the *nothing follows* response in Table 5 is an intuitive one in comparison to the $\neg C$ responses, they should be quicker, that is take lesser time. But according to the experimental data $\neg C$ responses took 3668 ms on average while *nothing follows* responses took 7151 ms. While the discussion about intuition and deliberation may be reserved for a later occasion, the WCS predicts that most reasoners respond with $\neg C$ for everyday conditional sentences during a DA inference task. In this case, most reasoners seem to (inherently) treat the antecedent of the conditional sentence as necessary, hence responding how they would in case of a bi-conditional sentence. Reasoners who upon deliberation have found a counterexample to the putative conclusion of $\neg A$ respond with *nothing follows*. This is the case, when these reasoners have comprehended the antecedent to be non-necessary. As mentioned in the beginning of Section 7, a non-necessary antecedent is one which allows the possibilities,

$$\begin{array}{cc} \neg A & \neg C \\ \neg A & C. \end{array}$$

A necessary antecedent on the other hand, disallows the latter. It must be pointed out that this also suggests that even if a reasoner deliberately searches for counterexamples to $\neg C$ they will be unable to find one, if they have comprehended the antecedent of the conditional as necessary.

The MMT predicts that in an AC task, given *if A then C* and *C*, most reasoners intuitively respond with *A*. When reasoners deliberate they respond with *nothing follows* in case they comprehend the conditional sentence as a conditional, meaning the possibilities,

$$\begin{array}{cc} A & C \\ \neg A & \neg C \\ \neg A & C. \end{array}$$

On the other hand, they stick to the putative response of A if they comprehend the conditional as a bi-conditional, meaning the possibilities,

$$\begin{array}{cc} A & C \\ \neg A & \neg C. \end{array}$$

Like the MMT, the WCS also predicts that most reasoners will answer A in the AC inference task. They seem to (inherently) treat the antecedents of conditional sentences as necessary; thereby treating the sentence as a bi-conditional. Furthermore, it predicts that when individuals look for counterexamples to their putative conclusion of A, they will sceptically respond *nothing follows*. In the search for counterexamples, the necessary or non-necessary nature of the antecedent with respect to the consequent seems to gain relevance, like in the DA task.

The MMT predicts that in the DC inference task, given *if A then C* and $\neg C$, most reasoners generally respond *nothing follows* as a result of intuition, and reasoners who respond $\neg A$ do so as a result of deliberation. This implies that the response *nothing follows* should be more rapid (take less time) than the response $\neg A$. However, the data from Table 7 in the Appendix, suggests that the mean response time for *nothing follows* responses is 5163 ms, whereas the mean response time for $\neg A$ responses is 4313 ms. The WCS on the other hand predicts that given an everyday conditional sentence most individuals may conclude $\neg A$. The response may plausibly be a result of the application of the *modus tollens* rule of inference as discussed in [8]. Upon deliberation, the previously discussed *factual* nature of the conditional may motivate the search for counterexamples which in turn may lead individuals to respond nothing follows. This prediction is limited to everyday conditionals where the antecedent and the consequent are related to each other to an acceptable or believable degree. Conditionals such as *if the sky is blue, then horses can speak English*, which may be considered bizarre, unacceptable or unbelievable are beyond the scope of the present discussion. However, it must also be acknowledged that, as Table 7 would tell the reader, the mean response time of 4558 ms for *nothing follows* responses in case of factual conditionals is comparable to that of 4595 ms for $\neg A$ responses. But a closer inspection of the conditionals (7) to (12) classified as factual, shows that in case of (8) and (11) *nothing follows* responses took longer than $\neg A$, whereas in case of (7) and (10) ¬A responses took longer than *nothing follows*. Conditionals (9) and (12) had comparable response times. It must be pointed out that it is plausible that each participant of the experiment might not have comprehended each conditional like the authors did. For example the conditional (6), *if it rains then the lawn must be wet*, had been classified as obligational by the authors but there might have been individuals who had deemed it to be factual. The ¬A response for this conditional had a median response time of 4062 ms, while the *nothing follows* response had 5235 ms. Overall, this motivates further research in part of the authors.

10 Conclusion

In this paper, we have discussed how the classification of conditional sentences and their antecedents help gain an insight into how humans understand or comprehend conditional sentences. On this basis, we have presented how the WCS along with its proposed extension can adequately model the average and the careful human reasoner in case of the DA inference task while also revisiting how the WCS can model the AC and the DC inference tasks.

In case of the DA, although most reasoners seem to respond with ¬C, the (necessary or non-necessary) type of the antecedent seems to be a relevant feature of the conditional sentence when a reasoner deliberates upon the task. As suggested by Philip Johnson-Laird in [18], given a set of premises, if one is beginning to form a conclusion, one should believe or adopt the same only if they are able to find no counterexamples strong enough to refute it. Table 5 in fact suggests that the reasoners responding *nothing follows* may actually be doing so, and such a response is due to the presence of counterexamples to their putative conclusion of ¬C. In case of the AC (like in the DA), reasoners who recognize the antecedent as non-necessary respond with *nothing follows*. In case of the DC, it is possibly the obligational or factual nature of the conditional sentence which is taken into consideration (see [8]) and reasoners with appropriate counterexamples respond *nothing follows*.

The case for the AA seems to be a ceiling effect, as an overwhelming majority of the responses were C (640 out of 672). The data has been omitted from the current discussion. Presently, it suffices to state that the WCS can well model this majority which indicates the possibility that during the AA task the conditional sentences were inherently taken to be obligational by most reasoners. This means, when A was affirmed they simply concluded C, probably due to the application of the *modus ponens* rule of inference. Nonetheless, although not significantly reflected in the current data for the AA, we do recognize that in case of factual conditionals where even if *A* holds, *C* may or may not, a reasoner might choose to respond *nothing follows*. Consider for example the conditional sentence, *if it is Monday, then Rita goes to school*. Given the factual nature of the conditional it seems plausible that sceptical reasoners may respond with *nothing follows*. WCS can also account for these reasoners, but it is outside the present scope of discussion. We believe the data at hand motivates further research about the AA and why humans accord with the response *C* so easily.

Returning to the DA task on the other hand, if we were to deny the antecedent, that is, *it is not Monday*, then although many reasoners might respond *Rita does not go to school*, once again, WCS with the extension proposed in this paper can account for such reasoners as well as those who choose to respond with scepticism that *nothing follows*.

11 Appendix

11.1 Conditionals used in the Experiment with Classification[3]

Obligational Conditionals with Necessary Antecedent (ON)

(1) *If it rains, then the roofs must be wet.*
(2) *If water in the cooking pot is heated over $99°C$, then the water starts boiling.*
(3) *If the wind is strong enough, then the sand is blowing over the dunes.*

Obligational Conditionals with Non-Necessary Antecedent (ONN)

(4) *If Paul rides a motorbike, then Paul must wear a helmet.*
(5) *If Maria is drinking alcoholic beverages in a pub, then Maria must be over 19 years*
    of age.
(6) *If it rains, then the lawn must be wet.*

Factual Conditionals with Necessary Antecedent (FN)

(7) *If the library is open, then Sabrina is studying late in the library.*
(8) *If the plants get water, then they will grow.*
(9) *If my car's start button is pushed, then the engine will start running.*

---

[3]Note: The classification was done by the authors of [7, 8].

Factual Conditionals with Non-Necessary Antecedent (FNN)

(10) *If Nancy rides her motorbike, then Nancy goes to the mountains.*
(11) *If Lisa plays on the beach, then Lisa will get sunburned.*
(12) *If Ron scores a goal, then Ron is happy.*

11.2 Short Background Story for Example 1

Maria and her friends are visiting a local pub to enjoy the evening with drinks and good food. Maria knows the local rules and regulations and obeys them.

11.3 Short Background Story for Example 2

The Presleys have moved into their newly built house and have hired a gardener to lay out the garden. They are sitting on their terrace and are looking at the bushes, small trees, and shrubs which were planted by the gardener two months ago.

11.4 Experiment Results for the AC Inference Task

Conditional/Classification	A	pct.	¬A	pct.	nf	pct.	Sum	Mdn A	Mdn nf
(1)	37		1		18		56	3952	7995
(2)	48		1		7		56	4003	4170
(3)	43		1		12		56	3458	9001
ON	128	76%	3	2%	37	22%	168	3804	7055
(4)	42		1		13		56	3659	8828
(5)	32		1		23		56	4704	6044
(6)	29		1		26		56	3593	4396
ONN	103	61%	3	2%	62	37%	168	3985	6423
(7)	51		1		4		56	3767	4397
(8)	42		1		13		56	3798	4565
(9)	45		1		10		56	3492	4598
FN	138	82%	3	2%	27	16%	168	3686	4520
(10)	34		2		20		56	5224	6289
(11)	29		2		25		56	3218	6205
(12)	33		1		22		56	3483	4992
FNN	96	57%	5	3%	67	40%	168	3975	5829
Obligational Conditional (O)	231	69%	6	2%	99	29%	336	3895	6739
Factual Conditional (F)	234	70%	8	2%	94	28%	336	3831	5175
Necessary Antecedent (N)	266	79%	6	2%	64	19%	336	3745	5788
Non-Necessary Antecedent (NN)	199	59%	8	2%	129	38%	336	3980	6126
Total	465	69%	14	2%	193	29%	672	3863	5957

Table 6: The results for AC inferences given a conditional sentence *if A then C* and an atomic fact C. In case of factual conditionals, *nf* is answered significantly more often. ON: obligational conditional with necessary antecedent, ONN: obligational conditional with non-necessary antecedent, FN: factual conditional with necessary antecedent, and FNN: factual conditional with non-necessary antecedent. All percentages (*pct.*) have been rounded off to the nearest natural number for the convenience of the reader.

11.5 Experiment Results for the DC Inference Task

Conditional/Classification	A	pct.	¬A	pct.	nf	pct.	Sum	$Mdn\ \neg A$	$Mdn\ nf$
(1)	1	2%	45	80%	10	18%	56	3449	4758
(2)	0	0%	50	89%	6	11%	56	4058	7922
(3)	2	4%	46	82%	8	14%	56	3796	4517
ON	3	2%	141	84%	24	14%	168	3768	5732
(4)	3	5%	46	82%	7	13%	56	3872	4154
(5)	1	2%	54	96%	1	2%	56	4946	8020
(6)	0	0%	36	64%	20	36%	56	4062	5235
ONN	4	2%	136	81%	28	17%	168	4293	5803
(7)	1	2%	37	66%	18	32%	56	5974	4744
(8)	3	5%	42	75%	11	20%	56	4367	5013
(9)	0	0%	47	84%	9	16%	56	4208	3966
FN	4	2%	126	75%	38	23%	168	4850	4574
(10)	2	4%	35	63%	19	34%	56	4879	4167
(11)	0	0%	39	70%	17	30%	56	4411	5647
(12)	0	0%	34	61%	22	39%	56	3726	3813
FNN	2	1%	108	64%	58	35%	168	4339	4542
Obligational Conditional (O)	7	2%	277	82%	52	15%	336	4031	5768
Factual Conditional (F)	6	2%	234	70%	96	29%	336	4595	4558
Necessary Antecedent (N)	7	2%	267	79%	62	18%	336	4309	5153
Non-Necessary Antecedent (NN)	6	2%	244	73%	86	26%	336	4316	5173
Total	13	2%	511	76%	148	22%	672	4313	5163

Table 7: The results for DC inferences given a conditional sentence *if A then C* and a negated atomic fact ¬C. In case of factual conditionals, *nf* is answered significantly more often. ON: obligational conditional with necessary antecedent, ONN: obligational conditional with non-necessary antecedent, FN: factual conditional with necessary antecedent, and FNN: factual conditional with non-necessary antecedent. All percentages (*pct.*) have been rounded off to the nearest natural number for the convenience of the reader.

References

[1] Krzysztof R. Apt and Maarten H. van Emden. Contributions to the theory of logic programming. *JACM*, 29:841–862, 1982.

[2] Meghna Bhadra. Comparative studies of the weak completion semantics and the mental model theory. Master's thesis, Technische Universität Dresden, Fakultät für Informatik, 2021.

[3] Ruth M. J. Byrne. Suppressing valid inferences with conditionals. *Cognition*, 31(1):61–83, 1989.

[4] Ruth M. J. Byrne. *The Rational Imagination: How People Create Alternatives to Reality*. MIT Press, Cambridge, MA, USA, 2005.

[5] Keith L. Clark. Negation as failure. In H. Gallaire and J. Minker, editors, *Logic and Databases*, pages 293–322. Plenum, New York, 1978.

[6] Kenneth J. W. Craik. *The Nature of Explanation*. Cambridge University Press, Cambridge, 1945.

[7] Marcos Cramer, Steffen Hölldobler, and Marco Ragni. Modeling human reasoning about conditionals. https://tu-dresden.de/ing/informatik/ki/krr/ressourcen/dateien/chr2021b.pdf/view, 2021. Accepted at NMR2021.

[8] Marcos Cramer, Steffen Hölldobler, and Marco Ragni. When are humans reasoning with modus tollens? *Proceedings of the Annual Conference of the Cognitive Science Society,*, 43:2337–2343, 2021. Retrieved from https://escholarship.org/uc/item/9x33q50g.

[9] Emmanuelle-Anna Dietz, Steffen Hölldobler, and Marco Ragni. A computational logic approach to the suppression task. *Proceedings of the Annual Conference of the Cognitive Science Society,*, 34:1500–1505, 2012. Retrieved from https://escholarship.org/uc/item/2sd6d61q.

[10] Christian Eichhorn, Gabriele Kern-Isberner, and Marco Ragni. Rational inference patterns based on conditional logic. In *Proceedings of the AAAI Conference on Artificial Intelligence*, volume 32, 2018.

[11] Melvin Fitting. A Kripke–Kleene semantics for logic programs. *Journal of Logic Programming*, 2(4):295–312, 1985.

[12] Melvin Fitting. *First-Order Logic and Automated Theorem Proving*. Springer-Verlag, Berlin, 2nd edition, 1996.

[13] Herbert Paul Grice. Logic and conversation. In P. Cole and J. L. Morgan, editors, *Syntax and Semantics*, volume 3, pages 41–58. Academic Press, New York, 1975.

[14] Islam Hamada and Steffen Hölldobler. On disjunctions and the weak completion semantics. In *Proceedings of the Virtual MathPsych/ICCM*. via mathpsych.org/presentation/571, 2021.

[15] Steffen Hölldobler. Weak completion semantics and its applications in human reasoning. In Ulrich Furbach and Claudia Schon, editors, *Bridging 2015 – Bridging the Gap between Human and Automated Reasoning*, volume 1412 of *CEUR Workshop Proceedings*, pages 2–16. CEUR-WS.org, 2015. http://ceur-ws.org/Vol-1412/.

[16] Steffen Hölldobler and Carroline D. P. Kencana Ramli. Logic programs under three-valued Łukasiewicz's semantics. In P. M. Hill and D. S. Warren, editors, *Logic Programming*, volume 5649 of *Lecture Notes in Computer Science*, pages 464–478. Springer-Verlag Berlin Heidelberg, 2009.

[17] Steffen Hölldobler and Carroline D. P. Kencana Ramli. Logics and networks for human reasoning. In C. Alippi, Marios M. Polycarpou, Christos G. Panayiotou, and Georgios Ellinasetal, editors, *Artificial Neural Networks – ICANN*, volume 5769 of *Lecture Notes*

in Computer Science, pages 85–94. Springer-Verlag Berlin Heidelberg, 2009.

[18] Philip N Johnson-Laird. Mental models in cognitive science. *Cognitive science*, 4(1):71–115, 1980.

[19] Philip N Johnson-Laird. Mental models and deduction. *Trends in cognitive sciences*, 5(10):434–442, 2001.

[20] Philip N. Johnson-Laird. Mental models and human reasoning. *Proceedings of the National Academy of Sciences*, 107(43):18243–18250, 2010.

[21] Philip N. Johnson-Laird and Ruth M. J. Byrne. *Deduction*. Lawrence Erlbaum Associates, Hove and London (UK), 1991.

[22] Philip N. Johnson-Laird and Ruth M. J. Byrne. Conditionals: A theory of meaning, pragmatics, and inference. *Psychological Review*, 109:646–678, 2002.

[23] Philip N Johnson-Laird, Geoffrey P Goodwin, and Sangeet S Khemlani. Mental models and reasoning. In *The Routledge International Handbook of Thinking and Reasoning*, pages 346–365. Routledge, 2017.

[24] Philip Nicholas Johnson-Laird. *Mental models: Towards a cognitive science of language, inference, and consciousness*. Number 6. Harvard University Press, 1983.

[25] PN Johnson-Laird. Models of deduction. *Reasoning: Representation and process in children and adults*, pages 7–54, 1975.

[26] PN Johnson-Laird and Sangeet S Khemlani. Toward a unified theory of reasoning. In *Psychology of learning and motivation*, volume 59, pages 1–42. Elsevier, 2013.

[27] Antonis C. Kakas, Robert A. Kowalski, and Francesca Toni. Abductive Logic Programming. *Journal of Logic and Computation*, 2(6):719–770, 1992.

[28] Carroline D. P. Kencana Ramli. Logic programs and three-valued consequence operators. Master's thesis, International Center for Computational Logic, TU Dresden, 2009.

[29] Sangeet S. Khemlani, Ruth M. J. Byrne, and Philip N. Johnson-Laird. Facts and possibilities: A model-based theory of sentenial reaoning. *Cognitive Science*, pages 1–38, 2018.

[30] Stephen C. Kleene. *Introduction to etamathematics*. North-Holland, 1952.

[31] John W. Lloyd. *Foundations of Logic Programming*. Springer-Verlag, 1984.

[32] Jan Łukasiewicz. O logice trójwartościowej. *Ruch Filozoficzny*, 5:169–171, 1920. English translation: On Three-Valued Logic. In: *Jan Łukasiewicz Selected Works*. (L. Borkowski, ed.), North Holland, 87-88, 1990.

[33] David Makinson. General patterns in nonmonotonic reasoning. handbook of logic in artificial intelligence and logic programming, vol. 3 (d. gabbay, c. hogger, j. robinson, editors), 1994.

[34] David Marr. Vision: A computational investigation into the human representation and processing of visual information, henry holt and co. *Inc., New York, NY*, 2(4.2), 1982.

[35] Ana Oliviera da Costa, Emmanuelle-Anna Dietz Saldanha, Steffen Hölldobler, and Marco Ragni. A computational logic approach to human syllogistic reasoning. *Proceedings of the Annual Conference of the Cognitive Science Society,*, 39:883–888, 2017.

[36] Wolfgang Spohn. Ordinal conditional functions: A dynamic theory of epistemic states. In *Causation in decision, belief change, and statistics*, pages 105–134. Springer, 1988.

[37] Wolfgang Spohn. *The laws of belief: Ranking theory and its philosophical applications*. Oxford University Press, 2012.

[38] Keith Stenning and Michiel van Lambalgen. Semantic interpretation as computation in nonmonotonic logic: The real meaning of the suppression task. *Cognitive Science*, 29:919–960, 2005.

[39] Keith Stenning and Michiel van Lambalgen. *Human Reasoning and Cognitive Science*. MIT Press, 2008.

Epistemic State Mappings among Ranking Functions and Total Preorders

Jonas Philipp Haldimann
FernUniversität in Hagen, Hagen, Germany
`jonas.haldimann@fernuni-hagen.de`

Christoph Beierle
FernUniversität in Hagen, Hagen, Germany
`christoph.beierle@fernuni-hagen.de`

Gabriele Kern-Isberner
University of Dortmund, Dortmund, Germany

Abstract

Ranking functions, also called ordinal conditional functions (OCFs), and total preorders on worlds (TPOs) are two common models for epistemic states that can represent conditional beliefs. To explore the connection between these frameworks, we consider mappings among TPOs and OCFs, i.e., the models of both frameworks. We formalize this kind of mappings as *epistemic state mappings*. Furthermore, we introduce postulates concerning the preservation of notable properties under the application of these mappings; a prominent example of such a property is syntax splitting. Other postulates regard the compatibility with operations like marginalization and conditionalization. We evaluate the interrelationships among the postulates for epistemic state mappings within and across the two frameworks, establishing dependencies as well as incompatibilities among postulates. Our results will be useful in particular for transferring methods and tools developed for OCF-based semantics to the TPO framework and the other way around.

1 Introduction

In the field of knowledge representation, there is a long tradition of employing conditionals as fundamental objects. A conditional formalizes a defeasible rule "If A

then usually B" for logical formulas A, B and is often denoted as $(B|A)$. As conditional logic is more expressive than propositional logic, it requires a richer semantics as well. There are different approaches to the semantics for conditional logic, e.g., [27, 1, 24, 29, 10, 6, 18]. These approaches often use either some form of ranking functions [33] or total preorders on interpretations as models for conditionals and conditional knowledge bases.

Both kinds of models, ranking functions (or ordinal conditional functions, OCFs) and total preorders on worlds (TPOs) have their own advantages. TPOs are fundamental for nonmonotonic logics and are used, e.g., in AGM revision [2] or the characterization of system P [1, 24]. OCFs are convenient implementations of TPOs that crucially provide the arithmetic that is lacking in TPOs. This arithmetic allows in particular for a more sophisticated conditional reasoning, approximating nicely what is possible in probabilistics. More specifically, TPOs are used in representation theorems for AGM revisions [17] and contractions [7] as well as system P inference [1, 24]. OCFs enable modelling the strength of conditional beliefs by assigning numbers to logical interpretations [33, 12]. Furthermore, some belief revision operators with interesting properties have been defined for OCFs, see e.g., c-changes [18]. To better understand the connection between OCFs and TPOs and to combine frameworks using these models, we investigate transformations among these frameworks, i.e., functions that map OCFs to TPOs or TPOs to OCFs, and as a generalization, we also take transformations from OCFs to OCFs and TPOs to TPOs into account.

By studying mappings between TPOs and OCFs and their properties, we expect insights that allow us to better transfer methods and tools developed for one of the frameworks to the other framework. E.g., the online system InfOCF-Web [26] and its underlying software library InfOCF-Lib [25] are currently capable of reasoning with ranking functions; it will be useful to extend them to cover reasoning with total preorders as well. Our results are also beneficial for a better connection of the postulates regarding syntax splitting in reasoning and belief revision on the different frameworks. An example of a successful transfer of a property for belief revision from one framework to another that has been made in the past is the definition of QPCP (qualitative principle of conditional preservation) for revisions of TPOs based on PCP (principle of conditional preservation) for the revision of OCFs [18, 19, 21].

Transferring belief change operations from one framework to the other framework is another possible application of this work. An approach to transferring operators across frameworks is to transfer an epistemic state of one kind (e.g. a TPO) into an epistemic state of the other kind (e.g. an OCF) and applying the operator in the new framework. This requires that the transformation conserves the properties of the epistemic state relevant to the operator.

We formalize functions on these models within and across the two different frameworks as *epistemic state mappings* and propose postulates that govern epistemic state mappings. The postulates require the epistemic state mappings to preserve certain properties of the models like the entailed inference relation and syntax splittings. Syntax splitting is a concept describing that beliefs about different parts of the signature are uncorrelated [28, 30]; it is used in postulates requiring that uncorrelated parts of the beliefs should be processed independently in revision and reasoning, see for example [28, 30, 22, 20]. Other postulates for epistemic state mappings ensure compatibility with the operations marginalization and conditionalization. These operations are relevant, e.g., for some forms of forgetting [9, 11, 5], syntax splitting, and certain aspects of belief revision [20, 31].

We investigate relationships among our postulates in general as well as for each framework in particular. Our results elaborate dependencies among the postulates, and they also unveil situations where certain combinations of postulates cannot be satisfied simultaneously. Especially for epistemic state mappings from TPOs to OCFs, there are several combinations of postulates that cannot be fulfilled simultaneously. Using the notion of coherence [23] we formulate weaker versions of our postulates that can be fulfilled at the same time.

In summary, the main contributions of this article are:

- Introduction of epistemic state mappings among TPOs and OCFs

- Coverage of marginalization and conditionalization also for the iterated case via the introduction of restricted TPOs and restricted OCFs

- Formalization of desirable properties of epistemic state mappings in terms of general postulates

- Establishment of relationships among the postulates and of realizability results for the postulates and for subsets thereof

- Postulates for epistemic state mappings from TPOs to OCFs based on the notion of coherence.

This article revises and extends our workshop submissions [13] and [15] and is structured as follows. After giving some background on conditional logic in Section 2, we introduce the operations marginalization and conditionalization and the property syntax splitting in Section 3. We introduce the concept of epistemic state mappings and postulates for such mappings in Section 4, and analyse the relationship among the postulates for different kinds of epistemic state mappings in Section 5. In Section 6 we propose weaker postulates based on coherence and show an epistemic state mapping fulfilling these postulates. In Section 7, we conclude and point out future work.

2 Background: Conditional Logic, Ranking Functions, and Total Preorders

A *(propositional) signature* is a finite set Σ of identifiers. For a signature Σ, we denote the propositional language over Σ by $\mathcal{L}_\Sigma$. Usually, we denote elements of the signatures with lowercase letters $a, b, c, \ldots$ and formulas with uppercase letters $A, B, C, \ldots$. We may denote a conjunction $A \wedge B$ by AB and a negation $\neg A$ by $\overline{A}$ for brevity of notation. As usual, $\top$ denotes a tautology and $\bot$ an unsatisfiable formula. The set of interpretations over a signature Σ is denoted as Ω_Σ. Interpretations are also called *worlds* and Ω_Σ is called the *universe*. An interpretation $\omega \in \Omega_\Sigma$ is a *model* of a formula $A \in \mathcal{L}_\Sigma$ if A holds in ω. This is denoted as $\omega \models A$. The set of models of a formula (over a signature Σ) is denoted as $Mod_\Sigma(A) = \{\omega \in \Omega_\Sigma \mid \omega \models A\}$. A formula A *entails* a formula B, denoted by $A \models B$, if $Mod_\Sigma(A) \subseteq Mod_\Sigma(B)$. We will represent interpretations (or worlds) by complete conjunctions, e.g., the interpretation over $\Sigma_{abc} = \{a, b, c\}$ that maps a and c to *true* and b to *false* is represented by $a \wedge \neg b \wedge c$, or just $a\overline{b}c$. Thus, every world $\omega \in \Omega_\Sigma$ is also a formula in $\mathcal{L}_\Sigma$.

A *conditional* $(B|A)$ connects two formulas A, B and represents the rule "If A then usually B". For a conditional $(B|A)$, the formula A is called the *antecedent* and the formula B the *consequent* of the conditional. The conditional language over a signature Σ is denoted as $(\mathcal{L}|\mathcal{L})_\Sigma = \{(B|A) \mid A, B \in \mathcal{L}_\Sigma\}$. $(\mathcal{L}|\mathcal{L})_\Sigma$ is a flat conditional language as it does not allow nesting conditionals. A finite set of conditionals is called a *conditional belief base*. We use a three-valued semantics of conditionals in this paper [8]. For a world ω, a conditional $(B|A)$ is either *verified* by ω if $\omega \models AB$, *falsified* by ω if $\omega \models A\overline{B}$, or *not applicable* to ω if $\omega \models \overline{A}$.

Conditionals are usually considered in the context of epistemic states. An *epistemic state* is a structure that represents all beliefs that are relevant for an agent's reasoning. There exist different kinds of models for epistemic states that can handle conditionals. Two approaches to this are ranking functions and total preorders on possible worlds.

A *ranking function*, also called *ordinal conditional function* (OCF), is a function $\kappa : \Omega_\Sigma \to \mathbb{N}_0$ such that $\kappa^{-1}(0) \neq \emptyset$; ranking functions were first introduced (in a more general form) by Spohn [33]. The intuition of a ranking function is that the rank of a world is lower if the world is more plausible. Therefore, ranking functions can be seen as some kind of "implausibility measure". For a ranking function κ and a set X of worlds, $\min_{\omega \in Mod(X)} \kappa(\omega)$ denotes the minimal rank $\kappa(\omega)$ among the worlds $\omega \in X$; for empty sets we define $\min_{\omega \in \emptyset} \kappa(\omega) = \infty$. Ranking functions are extended to formulas by $\kappa(A) = \min_{\omega \in Mod(A)} \kappa(\omega)$. A ranking function κ models a conditional $(B|A)$, denoted as $\kappa \models (B|A)$, if $\kappa(AB) < \kappa(A\overline{B})$, i.e., if the verification of the conditional is strictly more plausible than its falsification. A ranking function κ models a conditional belief base $\mathcal{R}$, denoted as $\kappa \models \mathcal{R}$ if $\kappa \models r$ for every $r \in \mathcal{R}$. The uniform ranking function κ_{uni} with $\kappa_{\text{uni}}(\omega) = 0$ for every $\omega \in \Omega_\Sigma$ represents the state where every possible world is equally plausible.

A *total preorder* (TPO) is a total, reflexive, and transitive binary relation. The meaning of a total preorder $\preceq$ on Ω_Σ as model for an epistemic state is that if $\omega_1 \preceq \omega_2$ then ω_1 is at least as plausible as ω_2 for $\omega_1, \omega_2 \in \Omega_\Sigma$. The strict version of a TPO $\preceq$ is the relation $\prec$ defined by $\omega_1 \prec \omega_2$ iff $\omega_1 \preceq \omega_2$ and $\omega_2 \not\preceq \omega_1$. For a TPO $\preceq$ and a set X of worlds, $\min(X, \preceq)$ denotes the set of minimal worlds in X with respect to $\preceq$. Total preorders on worlds are extended to formulas $A, B \in \mathcal{L}_\Sigma$ by defining $A \preceq B$ iff there is an $\omega_1 \in Mod_\Sigma(A)$ such that for every $\omega_2 \in Mod_\Sigma(B)$ it holds that $\omega_1 \preceq \omega_2$. A total preorder $\preceq$ models a conditional $(B|A)$, denoted as $\preceq \models (B|A)$, if $AB \prec A\overline{B}$, i.e., if the verification of the conditional is strictly more plausible than its falsification. A total preorder $\preceq$ models a conditional belief base $\mathcal{R}$, denoted as $\preceq \models \mathcal{R}$ if $\preceq \models (B|A)$ for every $(B|A) \in \mathcal{R}$.

The normalization requirement $\kappa^{-1}(0) \neq \emptyset$ for ranking functions ensures that the minimal worlds in a ranking function have rank 0. Therefore, worlds with rank 0 in a ranking function correspond to the minimal worlds in a total preorder: these worlds are the most plausible worlds in the epistemic state and determine the unconditional beliefs in this state. Such a normalization is not possible for total preorders; but it is also not necessary because their semantics use minimal worlds instead of worlds with rank 0.

While the definition of ranking functions used here does not allow for worlds to have rank ∞, all concepts and theorems developed in this article can be extended to cover also epistemic state representations comprising pieces of strict knowledge, i.e., representations where some worlds are considered to be impossible. The rank ∞ could be used to mark worlds that are impossible in OCFs; correspondingly, a set of worlds could be marked impossible in epistemic states based on TPOs.

Both ranking functions and total preorders each induce a total preorder on their domain. For a total preorder $\preceq$ the induced order $\leq_{\preceq}$ is the order $\preceq$ itself, i.e., $\leq_{\preceq} = \preceq$. For a ranking function κ, the induced ordering $\leq_\kappa$ is given by $\omega_1 \leq_\kappa \omega_2$ iff $\kappa(\omega_1) \leq \kappa(\omega_2)$ for $\omega_1, \omega_2 \in \Omega_\Sigma$.

3 Marginalization, Conditionalization, Syntax Splitting

We want to consider transformations among models of epistemic states represented by ranking functions or total preorders. To establish a notation for the domain of such transformations, we use the sets

$$\mathcal{M}_{TPO}(\Sigma) = \{\preceq \,\subseteq \Omega_\Sigma \times \Omega_\Sigma \mid \,\preceq \text{ is a total preorder over } \Omega_\Sigma\} \quad (1)$$
$$\mathcal{M}_{OCF}(\Sigma) = \{\kappa : \Omega_\Sigma \to \mathbb{N}_0 \mid \kappa \text{ is a ranking function}\} \quad (2)$$

containing all models over a certain signature Σ.

Marginalization and conditionalization are two basic operations in commonsense reasoning because they help intelligent agents to abstract from irrelevant features (marginalization) and focus on specific contexts (conditionalization). Technically, both operations help cutting down the complexity of reasoning and revision since they allow focusing on parts of the epistemic state.

Marginalization realizes focusing on certain signature elements; it is essential to, e.g., syntax splitting for epistemic states [22]. Conditionalization corresponds to focusing on a certain context or case [4]; it has applications, e.g., in belief revision and for ranking kinematics [31, 23, 32]. Both operations allow for concentrating on relevant parts of the beliefs, but in epistemically and technically different ways.

3.1 Marginalization

We start with defining marginalization in an abstract way.

Definition 1 (marginalization of ranking functions [33, 3]). *The marginalization of ranking functions from a signature Σ to a sub-signature $\Sigma' \subseteq \Sigma$ is a function $\mathcal{M}_{OCF}(\Sigma) \to \mathcal{M}_{OCF}(\Sigma')$, $\kappa \mapsto \kappa_{|\Sigma'}$ such that $\kappa_{|\Sigma'}(\omega) = \kappa(\omega)$ for $\omega \in \Omega_{\Sigma'}$.*

Definition 2 (marginalization of total preorders [3, 22]). *The marginalization of total preorders from signature Σ to a sub-signature $\Sigma' \subseteq \Sigma$ is a function $\mathcal{M}_{TPO}(\Sigma) \to \mathcal{M}_{TPO}(\Sigma')$, $\preceq \,\mapsto\, \preceq_{|\Sigma'}$ such that $\omega_1 \preceq_{|\Sigma'} \omega_2$ iff $\omega_1 \preceq \omega_2$ for $\omega_1, \omega_2 \in \Omega_{\Sigma'}$.*

The marginalizations of OCFs and TPOs presented above are special cases of general forgetful functors $Mod(\varrho)$ from Σ-models to Σ'-models given in [3] where $\Sigma' \subseteq \Sigma$ and $\varrho : \Sigma' \hookrightarrow \Sigma$ is the inclusion from Σ' to Σ. Informally, a forgetful functor forgets everything about the interpretation of the symbols in $\Sigma \setminus \Sigma'$ when mapping a Σ-model to a Σ'-model.

Marginalization can also be defined on formulas.

Definition 3 (marginalization of formulas). *Let Σ be a signature, $A \in \mathcal{L}_\Sigma$, and $\Sigma' \subseteq \Sigma$. For a world $\omega \in \Omega_\Sigma$ let $\omega_{\Sigma'} \in \Omega_{\Sigma'}$ be the assignment of truth values to variables in Σ' as in ω. A formula $A' \in \mathcal{L}_{\Sigma'}$ is a marginalization of A to Σ' if $Mod_{\Sigma'}(A') = \{\omega_{\Sigma'} \mid \omega \in Mod_\Sigma(A)\}$.*

If A' and A'' are marginalizations of A then $A' \equiv A''$. Thus, all marginalizations of A to Σ' are equivalent and we use $A_{|\Sigma'}$ to denote an arbitrary marginalization of A in situations where the specific syntax of the marginalization is not relevant.

3.2 Conditionalization

Conditionalization restricts the set of worlds that are considered in an epistemic state. After the conditionalization with a formula A, the resulting state only considers the worlds in $Mod_\Sigma(A)$ as possible worlds.

A conditionalization of OCFs was already introduced in [33], in analogy to probability theory. Both OCFs and probabilities make use of arithmetic operations to realize conditionalization; this, however, is also necessary because of the normalization condition which is not present for TPOs. Using tight transformations between TPOs and OCFs allows for transferring basic properties of OCF conditionalization and its arithmetics to the qualitative framework of OCFs. This helps elaborating non-numerical characteristics of conditionalization as a basic epistemic operation.

A notion of conditionalization for TPOs where the models of $\overline{A}$ are shifted to the uppermost layer has been introduced in [20]. Here, we will use the concept of conditionalization where the models of $\overline{A}$ are removed entirely from the epistemic state. To capture the outcome of such conditionalizations, we introduce the notions of restricted OCFs and TPOs.

Definition 4 (restricted OCF/TPO). *Let $M \subseteq \Omega_\Sigma$ be a set of worlds. A function $\kappa : M \to \mathbb{N}_0$ such that $\kappa^{-1}(0) \neq \emptyset$ is a restricted ranking function. A TPO $\preceq$ on M is called a restricted total preorder.*

Restricted ranking functions are extended to formulas by

$$\kappa(A) = \min\nolimits_{\omega \in (Mod_\Sigma(A) \cap M)} \kappa(\omega).$$

Restricted total preorders on worlds are extended to formulas by $A \preceq B$ iff there is an $\omega_1 \in Mod_\Sigma(A) \cap M$ such that for every $\omega_2 \in \min(Mod_\Sigma(B) \cap M)$ it holds that $\omega_1 \preceq \omega_2$.

The intuition of restricted OCFs and TPOs is the same as for OCFs and TPOs: Worlds with lower rank or position in the ordering are more plausible. For a signature Σ and a formula $A \in \mathcal{L}_\Sigma$ we define the set of restricted states

$$\mathcal{M}_{TPO}(\Sigma, A) = \{\preceq \, \subseteq Mod_\Sigma(A) \times Mod_\Sigma(A) \mid \, \preceq \text{ total preorder over } Mod_\Sigma(A)\} \tag{3}$$

$$\mathcal{M}_{OCF}(\Sigma, A) = \{\kappa : Mod_\Sigma(A) \to \mathbb{N}_0 \mid \kappa \text{ ranking function}\}. \tag{4}$$

The sets of restricted OCFs and TPOs properly include the sets of (unrestricted) OCFs and TPOs as given in (1) and (2) because $\mathcal{M}_I(\Sigma) = \mathcal{M}_I(\Sigma, \top)$ for $I \in \{TPO, OCF\}$. For a state $\Psi \in \mathcal{M}_I(\Sigma, A)$, we call $sig(\Psi) = \Sigma$ the signature of Ψ and $dom(\Psi) = Mod_\Sigma(A)$ the domain of Ψ. Using the concept of restricted OCFs, the coditionalization of OCFs adapted to our setting is the following.

Definition 5 (conditionalization of ranking functions [33, 31]). *The conditionalization of ranking functions over a signature Σ to the models of a formula $A \in \mathcal{L}_\Sigma$ is a function $\mathcal{M}_{OCF}(\Sigma) \to \mathcal{M}_{OCF}(\Sigma, A)$, $\kappa \mapsto \kappa|A$ such that $\kappa|A(\omega) = \kappa(\omega) - \kappa(A)$ for $\omega \in Mod_\Sigma(A)$.*

Likewise, we define conditionalization for TPOs such that the models of $\overline{A}$ are removed entirely from the TPO.

Definition 6 (conditionalization of total preorders). *The conditionalization of total preorders over a signature Σ to the models of a formula $A \in \mathcal{L}_\Sigma$ is a function $\mathcal{M}_{TPO}(\Sigma) \to \mathcal{M}_{TPO}(\Sigma, A)$, $\preceq \mapsto \preceq|A$ such that $\omega_1 \ (\preceq|A) \ \omega_2$ iff $\omega_1 \preceq \omega_2$ for $\omega_1, \omega_2 \in Mod_\Sigma(A)$.*

Note that Definitions 5 and 6 for conditionalization integrate nicely with our notions of restricted TPOs and restricted OCFs, because models of $\overline{A}$ occur neither in the elements of $\mathcal{M}_{OCF}(\Sigma, A)$ nor $\mathcal{M}_{TPO}(\Sigma, A)$.

3.3 Marginalization and Conditionalization on Restricted States

While originally, both marginalization and conditionalization were defined on OCFs and TPOs with the full set of Σ-models, we will also consider the iterative application of these operations. Therefore, we extend the definitions of these operations to cover already conditionalized states, which are restricted TPOs and OCFs.

Definition 7 (marginalization of restricted OCFs/TPOs). *Let Σ be a signature, $A \in \mathcal{L}_\Sigma$, and $\Sigma' \subseteq \Sigma$. The marginalization of restricted ranking functions over $Mod_\Sigma(A)$ from Σ to Σ' is a function $\mathcal{M}_{OCF}(\Sigma, A) \to \mathcal{M}_{OCF}(\Sigma', A_{|\Sigma'})$, $\kappa \mapsto \kappa_{|\Sigma'}$ such that $\kappa_{|\Sigma'}(\omega) = \kappa(\omega)$ for $\omega \in Mod_{\Sigma'}(A_{|\Sigma'})$.*

The marginalization of restricted total preorders over $Mod_\Sigma(A)$ from Σ to Σ' is a function $\mathcal{M}_{TPO}(\Sigma, A) \to \mathcal{M}_{TPO}(\Sigma', A_{|\Sigma'})$, $\preceq \mapsto \preceq_{|\Sigma'}$ such that $\omega_1 \preceq_{|\Sigma'} \omega_2$ iff $\omega_1 \preceq \omega_2$ for $\omega_1, \omega_2 \in Mod_{\Sigma'}(A_{|\Sigma'})$.

For ranking functions, the rank of a formula is not affected by marginalization if the formula only uses signature elements from the remaining subsignature.

Lemma 1. *Let κ be a (restricted) ranking function over Ω_Σ and A be a formula over $\Sigma' \subseteq \Sigma$. Then $\kappa(A) = \kappa_{|\Sigma'}(A)$.*

Definition 8 (conditionalization of restricted OCFs/TPOs). *Let Σ be a signature and A, B be formulas in $\mathcal{L}_\Sigma$. The conditionalization of restricted ranking functions over $Mod_\Sigma(B)$ to the models of A is a function $\mathcal{M}_{OCF}(\Sigma, B) \to \mathcal{M}_{OCF}(\Sigma, A \wedge B)$, $\kappa \mapsto \kappa|A$ such that $\kappa|A(\omega) = \kappa(\omega) - \kappa(A)$ for $\omega \in Mod_\Sigma(A \wedge B)$.*

3	$\bar{a}\bar{b}c$		3			3	
2	$\bar{a}bc$ $\bar{a}b\bar{c}$		2	$\bar{a}$		2	
1	$a\bar{b}c$		1			1	$a\bar{b}c$
0	abc $ab\bar{c}$		0	a		0	abc

(a) A restricted OCF κ with the syntax splitting $\{\{a\}, \{b, c\}\}$.

(b) Marginalization $\kappa_{|\{a\}}$ of κ.

(c) Conditionalization $\kappa|\{ac\}$ of κ.

Figure 1: Example for syntax splitting, marginalization, and conditionalization on restricted OCFs.

The conditionalization of restricted total preorders over $Mod_\Sigma(B)$ *to the models of* A *is a function* $\mathcal{M}_{TPO}(\Sigma, B) \to \mathcal{M}_{TPO}(\Sigma, A \wedge B)$, $\preceq \mapsto \preceq|A$ *such that* $\omega_1 \preceq|A\ \omega_2$ *iff* $\omega_1 \preceq \omega_2$ *for* $\omega_1, \omega_2 \in Mod_\Sigma(A \wedge B)$.

For $\mathcal{M}_I(\Sigma, \top)$, the marginalization/conditionalization of the restricted OCFs/TPOs coincides with the marginalization/conditionalization of OCFs/TPOs. Thus, Definitions 7 and 8 of marginalization and conditionalization properly cover and extend the Definitions 1, 2, 5, and 6.

Example 1. *Consider the restricted ranking function* κ *in Figure 1a. The marginalisation* $\kappa_{|\{a\}}$ *of* κ *is shown in Figure 1b and the conditionalization* $\kappa|a$ *of* κ *is shown in Figure 1c.*

3.4 Syntax Splitting for Epistemic States

An interesting feature of ranking functions and total preorders is the existence of syntax splittings. Syntax splittings were first introduced as a property of belief sets by Parikh [28]. Informally, the meaning of a belief set having a syntax splitting is that the belief set contains independent information over different parts of the signature. The partition of the signature in these parts is called a syntax splitting for the considered belief set. Syntax splittings are useful properties as they indicate that different parts of the belief state should be processed independently of each other. This can be used to formulate postulates for sensible reasoning and revision operators. Additionally, splitting belief states and processing their parts independently can make operations computationally more efficient.

The notion of syntax splitting was extended to other representations of epistemic states such as TPOs and OCFs in [22]. For a partitioning $\{\Sigma_1, \ldots, \Sigma_n\}$ of a signature

Σ and a world $\omega \in \Omega_\Sigma$, let the world $\omega^i \in \Omega_{\Sigma_i}$ denote the variable assignment of the variables in Σ_i as in ω in the following definitions. Let the world $\omega^{\neq i} \in \Omega_{\Sigma \setminus \Sigma_i}$ denote the variable assignment of the variables in $\Sigma \setminus \Sigma_i$ as in ω.

Definition 9 (syntax splitting for total preorders [22]). *Let $\preceq$ be a total preorder over a signature Σ. A partitioning $\{\Sigma_1, \ldots, \Sigma_n\}$ of Σ is a syntax splitting for $\preceq$ if, for $i = 1, \ldots, n$,*

$$\omega_1^{\neq i} = \omega_2^{\neq i} \quad \text{implies} \quad (\omega_1 \preceq \omega_2 \text{ iff } \omega_1^i \preceq_{|\Sigma_i} \omega_2^i).$$

Definition 10 (syntax splitting for ranking functions [22]). *Let κ be a ranking function over Σ. A partitioning $\{\Sigma_1, \ldots, \Sigma_n\}$ of Σ is a syntax splitting for κ if there are ranking functions $\kappa_i : \Sigma_i \to \mathbb{N}_0 \cup \{\infty\}$ for $i = 1, \ldots, n$ such that $\kappa(\omega) = \kappa_1(\omega^1) + \cdots + \kappa_n(\omega^n)$. This is denoted as $\kappa = \kappa_1 \oplus \cdots \oplus \kappa_n$.*

The notion of syntax splitting can be extended to restricted OCFs and TPOs.

Definition 11 (syntax splitting for restricted TPOs). *Let $A \in \mathcal{L}_\Sigma$ be a formula and $\preceq$ be a restricted total preorder in $\mathcal{M}_{TPO}(\Sigma, A)$. A partitioning $\{\Sigma_1, \ldots, \Sigma_n\}$ of Σ is a syntax splitting for $\preceq$ if*

- *there are formulas $A_1, \ldots, A_n$ such that $A \equiv A_1 \wedge \cdots \wedge A_n$ and $A_i \in \mathcal{L}_{\Sigma_i}$ for $i = 1, \ldots, n$*

- *and, for $i = 1, \ldots, n$ and $\omega_1, \omega_2 \in dom(\preceq)$,*

$$\omega_1^{\neq i} = \omega_2^{\neq i} \quad \text{implies} \quad (\omega_1 \preceq \omega_2 \text{ iff } \omega_1^i \preceq_{|\Sigma_i} \omega_2^i).$$

Definition 12 (syntax splitting for restricted OCFs). *Let $A \in \mathcal{L}_\Sigma$ be a formula and κ be a restricted ranking function in $\mathcal{M}_{OCF}(\Sigma, A)$. A partitioning $\{\Sigma_1, \ldots, \Sigma_n\}$ of Σ is a syntax splitting for κ if*

- *there are formulas $A_1, \ldots, A_n$ such that $A \equiv A_1 \wedge \cdots \wedge A_n$ and $A_i \in \mathcal{L}_{\Sigma_i}$ for $i = 1, \ldots, n$*

- *and there are ranking functions $\kappa_i \in \mathcal{M}_{OCF}(\Sigma_i, A_i)$ for $i = 1, \ldots, n$ such that $\kappa(\omega) = \kappa_1(\omega^1) + \cdots + \kappa_n(\omega^n)$ for $\omega \in dom(\kappa)$.*

This is denoted as $\kappa = \kappa_1 \oplus \cdots \oplus \kappa_n$.

Again, the definitions of syntax splitting for restricted total preorders or ranking functions cover the definitions of syntax splitting for TPOs/OCFs: A partition of Σ is a syntax splitting for a total preorder $\preceq \in \mathcal{M}_{TPO}(\Sigma)$ according to Definition 9 iff it is a syntax splitting for $\preceq$ according to Definition 11. A partition is a syntax splitting for a ranking function $\kappa \in \mathcal{M}_{OCF}(\Sigma)$ according to Definition 10 iff it is a syntax splitting for κ according to Definition 12.

Note that a syntax splitting of an OCF κ is also a syntax splitting for the TPO $\leqslant_\kappa$ induced by κ, but not the other way round (see [22]).

Example 2. *The restricted ranking function κ over $\Sigma = \{a,b,c\}$ displayed in Figure 1a and the total preorder induced by κ both have the syntax splitting $\{\{a\},\{b,c\}\}$.*

4 Postulates for Mappings on Epistemic States

TPOs and OCFs have each their own advantages. While calculating with plausibilities is easier in an OCF-based framework, TPOs are more closely linked to qualitative beliefs. To work with both frameworks and to transfer approaches from one framework to the other, we are interested in transformations from OCFs to TPOs and vice versa. More generally, we want to investigate transformations among epistemic states represented by TPOs or by OCFs.

We formalize the transformations among models of epistemic states by introducing so-called epistemic state mappings. To capture that an epistemic state mapping should cover transformations for different signatures, we define epistemic state mappings as function families providing a mapping for every signature and every set of worlds over this signature. Note that *TPO* and *OCF* are used as symbols representing the type of an epistemic state in the following definition.

Definition 13. *Let $I_1, I_2 \in \{TPO, OCF\}$. An epistemic state mapping from I_1 to I_2, denoted as $\xi : I_1 \leadsto I_2$, is a function family $\xi = (\xi_{\Sigma,A})$ for signatures Σ and formulas $A \in \mathcal{L}_\Sigma$ with $\xi_{\Sigma,A} : \mathcal{M}_{I_1}(\Sigma, A) \to \mathcal{M}_{I_2}(\Sigma, A)$ such that $A \equiv B$ implies $\xi_{\Sigma,A} = \xi_{\Sigma,B}$.*

Definition 13 covers four types of epistemic state mappings. A mapping $\xi_1 : TPO \leadsto TPO$ maps total preorders to total preorders; a mapping $\xi_2 : TPO \leadsto OCF$ maps total preorders to ranking functions; a mapping $\xi_3 : OCF \leadsto OCF$ maps ranking functions to ranking functions; and a mapping $\xi_4 : OCF \leadsto TPO$ maps ranking functions to total preorders. Applying an epistemic state mapping always leaves the domain of the epistemic state unchanged, e.g., for a ranking function $\kappa : Mod_\Sigma(A) \to \mathbb{N}_0$ we have $dom(\xi_{3\ \Sigma,A}(\kappa)) = Mod_\Sigma(A)$.

Note that the last requirement for epistemic state mappings given in Definition 13 amounts to syntax independence for index A of the function family $(\xi_{\Sigma,A})$.

Example 3. *Consider the family of functions $\xi^{reverse} : TPO \rightsquigarrow TPO$ defined by, for a signature Σ, a formula $A \in \mathcal{L}_\Sigma$, and $\preceq \, \in \mathcal{M}_{TPO}(\Sigma, A)$,*

$$\xi^{reverse}_{\Sigma,A}(\preceq) = \preceq' \quad \text{with} \quad \omega_1 \preceq' \omega_2 \text{ iff } \omega_2 \preceq \omega_1 \quad \text{for } \omega_1, \omega_2 \in Mod_\Sigma(A).$$

The function family $\xi^{reverse}$ is an epistemic state mapping from TPOs to TPOs that reverses a TPO. The family of functions τ^* defined by, for a signature Σ, a formula $A \in \mathcal{L}_\Sigma$, and $\kappa \in \mathcal{M}_{OCF}(\Sigma, A)$,

$$\tau^*_{\Sigma,A}(\kappa) = \preceq_\kappa$$

is an epistemic state mapping from TPOs to OCFs, that maps every OCF to the TPO induced by it.

Every epistemic state mapping represents a way to transform epistemic states of kind I_1 to epistemic states of kind I_2 for different domains. Desirable properties of epistemic state mappings ($\xi_{\Sigma,A}$) can be stated in the form of postulates. For instance, regarding the conditional inference represented by an epistemic state, we could require that the set of accepted conditional beliefs should not be reduced, should not be increased, or should be kept identical.

Postulates. *Let $I_1, I_2 \in \{TPO, OCF\}$ and let $\xi : I_1 \rightsquigarrow I_2$ be an epistemic state mapping from I_1 to I_2. Let Σ be a signature, $A \in \mathcal{L}_\Sigma$, and $\Psi \in \mathcal{M}_{I_1}(\Sigma, A)$.*

Let $(C|D) \in (\mathcal{L} \mid \mathcal{L})_\Sigma$.

(IE) $\quad \Psi \models (C|D) \quad \text{iff} \quad \xi_{\Sigma,A}(\Psi) \models (C|D)$.

(wIE$^\Rightarrow$) $\quad \Psi \models (C|D) \quad \text{implies} \quad \xi_{\Sigma,A}(\Psi) \models (C|D)$.

(wIE$^\Leftarrow$) $\quad \xi_{\Sigma,A}(\Psi) \models (C|D) \quad \text{implies} \quad \Psi \models (C|D)$.

The postulate (IE) requires *inferential equivalence* and states that the epistemic state mapping may not change the set of conditionals accepted by an epistemic state. The epistemic state and its mapping induce the same inference relation with respect to conditionals. This is quite a strong postulate, the postulates (wIE$^\Rightarrow$) and (wIE$^\Leftarrow$) are weaker versions of (IE). Postulate (wIE$^\Rightarrow$) states that an epistemic state mapping may not remove conditionals from the set of inferred conditionals. Postulate (wIE$^\Leftarrow$) states that after applying an epistemic state mapping, we may not accept additional conditionals.

Postulates. *Let $I_1, I_2 \in \{TPO, OCF\}$ and let $\xi : I_1 \leadsto I_2$ be an epistemic state mapping. Let Σ be a signature, $A \in \mathcal{L}_\Sigma$, and $\Psi \in \mathcal{M}_{I_1}(\Sigma, A)$.*

Let ω_1, ω_2 in $dom(\Psi)$.

(Ord) $\quad \omega_1 <_\Psi \omega_2 \quad$ iff $\quad \omega_1 <_{\xi_{\Sigma, A}(\Psi)} \omega_2$.

(wOrd$^\Rightarrow$) $\quad \omega_1 <_\Psi \omega_2 \quad$ implies $\quad \omega_1 <_{\xi_{\Sigma, A}(\Psi)} \omega_2$

(wOrd$^\Leftarrow$) $\quad \omega_1 <_{\xi_{\Sigma, A}(\Psi)} \omega_2 \quad$ implies $\quad \omega_1 <_\Psi \omega_2$.

The postulate (Ord) states that the ordering on worlds induced by an epistemic state should not be changed by the epistemic state mapping, i.e., if a world ω_1 is more plausible than a world ω_2 in a TPO or OCF Ψ, then ω_1 should be more plausible than ω_2 in $\xi(\Psi)$ as well. While (Ord) is quite a strong postulate, the weaker versions (wOrd$^\Rightarrow$) and (wOrd$^\Leftarrow$) cover only one direction of the "iff" in (Ord). (wOrd$^\Rightarrow$) states that if a world ω_1 is more plausible than another world ω_2 before applying the epistemic state mapping, then ω_1 should still be more plausible than ω_2 after applying the epistemic state mapping. (wOrd$^\Leftarrow$) requires that a world ω_1 can only be more plausible than another world ω_2 after applying the epistemic state mapping, if ω_1 was already more plausible than ω_2 before the application of the epistemic state mapping.

To some extent, postulate (Ord) can be seen as a generalization of the idea of kinematics in probabilistics and for ranking functions [32] in the sense that it requires the preservation of relations between worlds after the application of an operation – conditionalization for probabilities and ranking functions, epistemic state mappings here.

It is easy to see that (IE) is equivalent to the conjunction of (wIE$^\Rightarrow$) and (wIE$^\Leftarrow$) and that (Ord) is equivalent to the conjunction of (wOrd$^\Rightarrow$) and (wOrd$^\Leftarrow$). Other relationships among the postulates are less obvious.

Proposition 1. *The following relationships hold between the introduced postulates:*

1. *(IE) is equivalent to (Ord).*
2. *(wIE$^\Rightarrow$) is equivalent to (wOrd$^\Rightarrow$).*
3. *(wIE$^\Leftarrow$) is equivalent to (wOrd$^\Leftarrow$).*

Proof. Let $\xi : I_1 \leadsto I_2$ be a epistemic state mapping with $I_1, I_2 \in \{TPO, OCF\}$.
Ad (2): "$\Leftarrow$" Let $(\xi_{\Sigma, A})$ satisfy (wOrd$^\Rightarrow$). Let $\Psi \in \mathcal{M}_{I_1}(\Sigma, A)$ and $\Phi = \xi_{\Sigma, A}(\Psi)$. If $\Psi \models (D|C)$, then $\min(Mod_\Sigma(CD), <_\Psi) <_\Psi \min(Mod_\Sigma(C\overline{D}), <_\Psi)$. In

this case, (wOrd$^\Rightarrow$) implies $\min(Mod_\Sigma(CD), <_\Phi) <_\Phi \min(Mod_\Sigma(C\overline{D}), <_\Phi)$. This is equivalent to $\Phi \models (C|D)$. Therefore, $(\xi_{\Sigma,A})$ satisfies (wIE$^\Rightarrow$).

"$\Rightarrow$" Let $(\xi_{\Sigma,A})$ satisfy (wIE$^\Rightarrow$). Let $\Psi \in \mathcal{M}_{I_1}(\Sigma, A)$ and $\Phi = \xi_{\Sigma,A}(\Psi)$. Let $\omega_1, \omega_2 \in \Omega$ with $\omega_1 <_\Psi \omega_2$. Then, $\Psi \models (\omega_1|\omega_1 \vee \omega_2)$. (wIE$^\Rightarrow$) implies that $\Phi \models (\omega_1|\omega_1 \vee \omega_2)$. Therefore, $\omega_1 <_\Phi \omega_2$. We see that $(\xi_{\Sigma,A})$ satisfies (wOrd$^\Rightarrow$).

Ad (3): "$\Leftarrow$" Let $(\xi_{\Sigma,A})$ satisfy (wOrd$^\Leftarrow$). Let $\Psi \in \mathcal{M}_{I_1}(\Sigma, A)$ and $\Phi = \xi_{\Sigma,A}(\Psi)$. If $\Phi \models (D|C)$, then $\min(Mod_\Sigma(CD), <_\Phi) <_\Phi \min(Mod_\Sigma(C\overline{D}), <_\Phi)$. In this case, (wOrd$^\Leftarrow$) implies $\min(Mod_\Sigma(CD), <_\Psi) <_\Psi \min(Mod_\Sigma(C\overline{D}), <_\Psi)$. This is equivalent to $\Psi \models (D|C)$. Therefore, $(\xi_{\Sigma,A})$ satisfies (wIE$^\Leftarrow$).

"$\Rightarrow$" Let $(\xi_{\Sigma,A})$ satisfy (wIE$^\Leftarrow$). Let $\Psi \in \mathcal{M}_{I_1}(\Sigma, A)$ and $\Phi = \xi_{\Sigma,A}(\Psi)$. Let $\omega_1, \omega_2 \in \Omega$ with $\omega_1 <_\Phi \omega_2$. Then, $\Phi \models (\omega_1|\omega_1 \vee \omega_2)$. (wIE$^\Leftarrow$) implies that $\Psi \models (\omega_1|\omega_1 \vee \omega_2)$. Therefore, $\omega_1 <_\Psi \omega_2$. We see that $(\xi_{\Sigma,A})$ satisfies (wOrd$^\Leftarrow$).

Ad (1): This follows from (2) and (3) as (IE) is the conjunction of (wIE$^\Rightarrow$) and (wIE$^\Leftarrow$) and (Ord) is the conjunction of (wOrd$^\Rightarrow$) and (wOrd$^\Leftarrow$). □

Postulates for revision and contraction of total preorders and ranking functions with syntax splitting (as in Definitions 9 and 10) have been introduced and investigated in [22], [16], and [14]. Because syntax splittings are a significant property of epistemic states, it is of interest to identify epistemic state mappings that preserve them.

Postulates. Let $I_1, I_2 \in \{TPO, OCF\}$ and let $\xi : I_1 \rightsquigarrow I_2$ be an epistemic state mapping. Let Σ be a signature, $A \in \mathcal{L}_\Sigma$, and $\Psi \in \mathcal{M}_{I_1}(\Sigma, A)$.

(SynSplit) If $\{\Sigma_1, \ldots, \Sigma_n\}$ is a syntax splitting for Ψ, then $\{\Sigma_1, \ldots, \Sigma_n\}$ is a syntax splitting for $\xi_{\Sigma,A}(\Psi)$.

(SynSplit^b) If $\{\Sigma_1, \Sigma_2\}$ is a syntax splitting for Ψ, then $\{\Sigma_1, \Sigma_2\}$ is a syntax splitting for $\xi_{\Sigma,A}(\Psi)$.

(SynSplit) states that an epistemic state mapping preserves syntax splittings of the epistemic state. In the literature addressing syntax splittings, the focus is sometimes on binary syntax splittings [28, 30, 20]; (SynSplit^b) is a splitting postulate for the special case of syntax splittings in two subsignatures.

Depending on the context in which we want to use the epistemic state mappings, compatibility with marginalization and conditionalization might be useful.

Postulates. Let $I_1, I_2 \in \{TPO, OCF\}$ and let $\xi : I_1 \rightsquigarrow I_2$ be an epistemic state mapping. Let Σ be a signature, $A \in \mathcal{L}_\Sigma$, and $\Psi \in \mathcal{M}_{I_1}(\Sigma, A)$.

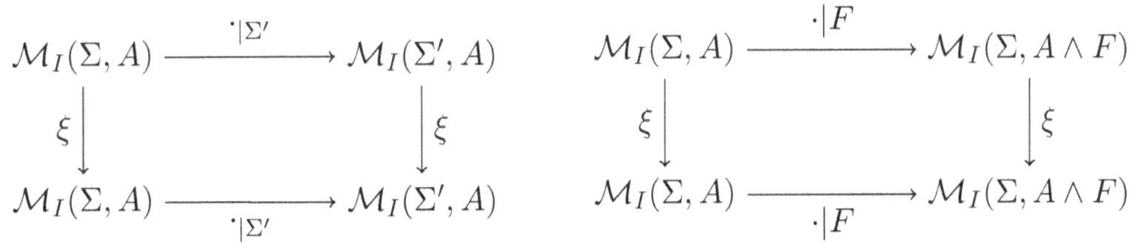

$$\begin{array}{ccc}
\mathcal{M}_I(\Sigma, A) \xrightarrow{\cdot|_{\Sigma'}} \mathcal{M}_I(\Sigma', A) & \quad & \mathcal{M}_I(\Sigma, A) \xrightarrow{\cdot|F} \mathcal{M}_I(\Sigma, A \wedge F) \\
\xi \downarrow \qquad \qquad \downarrow \xi & & \xi \downarrow \qquad \qquad \downarrow \xi \\
\mathcal{M}_I(\Sigma, A) \xrightarrow[\cdot|_{\Sigma'}]{} \mathcal{M}_I(\Sigma', A) & & \mathcal{M}_I(\Sigma, A) \xrightarrow[\cdot|F]{} \mathcal{M}_I(\Sigma, A \wedge F)
\end{array}$$

(a) Illustration of (Marg).   (b) Illustration of (Cond).

Figure 2: Commutative diagrams illustrating the postulates (Cond) and (Marg).

Let $\Sigma' \subseteq \Sigma$ with $\Sigma' \neq \emptyset$ and $A' = A_{|\Sigma'}$.

(Marg) $\xi_{\Sigma', A'}(\Psi_{|\Sigma'}) = (\xi_{\Sigma, A}(\Psi))_{|\Sigma'}$

Let $F \in \mathcal{L}_\Sigma$ with $\mathrm{Mod}_\Sigma(F) \cap \mathrm{dom}(\Psi) \neq \emptyset$.

(Cond) $\xi_{\Sigma, A \wedge F}(\Psi|F) = (\xi_{\Sigma, A}(\Psi))|F$

The postulate (Marg) ensures the compatibility of an epistemic state mapping with marginalization. It states that changing the order in which marginalization and the epistemic state mapping are applied does not matter. This postulate is illustrated in Figure 2a. Similarly, the postulate (Cond) ensures the compatibility of an epistemic state mapping with conditionalization. (Cond) is illustrated in Figure 2b.

In the next sections, we will investigate the introduced postulates further for specific combinations of I_1 and I_2.

5 Epistemic State Mappings of Different Types

In the following, we investigate epistemic state mappings from TPOs to TPOs (Section 5.1), OCFs to OCFs (Section 5.2), OCFs to TPOs (Section 5.3), and TPOs to OCFs (Section 5.4) in more detail.

5.1 Mapping Total Preorders to Total Preorders

Let us first consider epistemic state mappings from total preorders to total preorders. If we want (IE) or the equivalent (Ord) to hold, we do not have much choice.

Proposition 2. *The only epistemic state mapping from TPOs to TPOs that fulfils (Ord) is the identity.*

From Proposition 2 it follows that (IE) or (Ord) imply (SynSplit), (Cond), and (Marg) for epistemic state mappings from TPOs to TPOs as the identity fulfils these postulates.

Proposition 3. *Let $\xi : TPO \rightsquigarrow TPO$ be an epistemic state mapping from TPOs to TPOs. If ξ fulfils (Ord) or (IE) then ξ fulfils (SynSplit), (Cond), and (Marg).*

The situation changes if we require only (wOrd$^\Rightarrow$) or (wOrd$^\Leftarrow$). We can think of a total preorder $\preceq$ as a stack of "layers". We say that two worlds $\omega_1, \omega_2 \in dom(\preceq)$ have the same position in $\preceq$, denoted as $\omega_1 \approx_\preceq \omega_2$, if $\omega_1 \preceq \omega_2$ and $\omega_1 \preceq \omega_2$. The relation $\approx_\preceq$ is an equivalence relation and layers are the equivalence classes of $\approx_\preceq$ on $dom(\preceq)$. I.e., two worlds $\omega_1, \omega_2 \in dom(\preceq)$ are in the same layer if they have the same position in the TPO. The layers are stacked according to the TPO: the lower a layer is, the smaller the worlds in it are with respect to $\preceq$.

A consequence of each (wOrd$^\Rightarrow$) and (wOrd$^\Leftarrow$) is that we cannot swap parts of different layers. If $\omega_1 \prec \omega_2$ for worlds ω_1, ω_2 then it is not possible that $\omega_2 \prec' \omega_1$ with $\preceq' = \xi(\preceq)$ if ξ fulfils either (wOrd$^\Rightarrow$) or (wOrd$^\Leftarrow$).

(wOrd$^\Rightarrow$) allows the "splitting" of layers. For worlds ω_1, ω_2 with $\omega_1 \approx_\preceq \omega_2$ we may have an epistemic state mapping ξ fulfilling (wOrd$^\Rightarrow$) with $\omega_1 \prec' \omega_2$ where $\preceq' = \xi(\preceq)$. Thus, (wOrd$^\Rightarrow$) allows to extend the set of accepted conditionals as stated in the equivalent postulate (wIE$^\Rightarrow$). However, the opposite is not allowed: An epistemic state mapping fulfilling (wOrd$^\Rightarrow$) may not merge parts of different layers together.

With respect to (wOrd$^\Leftarrow$) we obtain dual observations. For ω_1, ω_2 with $\omega_1 \prec \omega_2$ we may have an epistemic state mapping ξ fulfilling (wOrd$^\Leftarrow$) with $\omega_1 \approx_{\xi(\preceq)} \omega_2$, i.e., merging of layers is allowed. Note that in this case we have $\omega_1 \approx_{\xi(\preceq)} \omega_3 \approx_{\xi(\preceq)} \omega_2$ for any ω_3 with $\omega_1 \preceq \omega_3 \preceq \omega_2$. Thus, (wOrd$^\Leftarrow$) allows us to reduce the set of accepted conditionals (as stated in (wIE$^\Leftarrow$)) which can be seen as a form of forgetting [5]. However, (wOrd$^\Leftarrow$) does not allow splitting of layers.

5.2 Mapping Ranking Functions to Ranking Functions

Let us consider the case where we map ranking functions to ranking functions. For this case there are epistemic state mappings fulfilling the postulates (IE), (Ord), (SynSplit), (Cond), and (Marg) simultaneously.

Proposition 4. *Let $a \in \mathbb{N}^+$. The epistemic state mapping $\xi : OCF \rightsquigarrow OCF, \kappa \mapsto a \cdot \kappa$ fulfils (IE), (Ord), (SynSplit), (Cond), and (Marg).*

We again use the concept of layers introduced in Section 5.1. For a ranking function κ, each layer contains the worlds in $\kappa^{-1}(k)$ for a certain $k \in \mathbb{N}_0$. Contrary to total preorders, ranking functions can have empty layers. These empty layers (or the lack thereof) make ranking functions more expressive than total preorders.

The implications of (Ord) for epistemic state mappings from OCFs to OCFs are similar to the implications for epistemic state mappings from TPOs to TPOs in terms of layers. The layers are not swapped, split, or merged by the epistemic state mapping. However, (Ord) allows for adding or removing empty layers. For example, the epistemic state mapping that removes all empty layers beneath a non-empty layer fulfils (Ord).

In contrast to (Ord), the postulate (wOrd$^\Rightarrow$) allows splitting of layers. If two worlds have the same rank in a ranking function κ they may have different ranks in $\xi(\kappa)$ without violating (wOrd$^\Rightarrow$). But (wOrd$^\Rightarrow$) prevents merging different layers. If two worlds have different ranks in a ranking function κ before the epistemic state mapping, they may not have the same rank in $\xi(\kappa)$.

The postulate (wOrd$^\Leftarrow$) allows merging but not splitting of layers. If two worlds ω_1, ω_2 have different ranks in κ they may have the same rank in $\kappa' = \xi(\kappa)$ without violating (wOrd$^\Leftarrow$). In this case it holds that $\kappa'(\omega_1) = \kappa'(\omega_2) = \kappa'(\omega_3)$ for any world ω_3 with $\kappa(\omega_1) \leqslant \kappa(\omega_3) \leqslant \kappa(\omega_2)$.

The next proposition provides a strong representation result, because it shows that the epistemic state mappings used in Proposition 4 are precisely the epistemic state mappings from OCFs to OCFs fulfilling the postulates (Ord), (Cond), and (Marg).

Proposition 5. *Let $\xi : OCF \rightsquigarrow OCF$ be an epistemic state mapping from OCFs to OCFs fulfilling (Ord), (Cond), and (Marg). Then there is an $a \in \mathbb{N}^+$ such that for every (restricted) ranking function κ it holds that $\xi(\kappa) = a \cdot \kappa$.*

Proof. Let $\xi : OCF \rightsquigarrow OCF$ be an epistemic state mapping. We need to show that there is an $a \in \mathbb{N}$ such that

$$\xi_{\Sigma,A}(\kappa) = a \cdot \kappa \tag{5}$$

for any Σ and $A \in \mathcal{L}_\Sigma$ and $\kappa : Mod_\Sigma(A) \to \mathbb{N}_0$.

We do this in three parts. First, we consider some arbitrary signature Σ with $|\Sigma| \geqslant 3$. We show that there is an a such that for any ranking function $\kappa_1 \in \mathcal{M}_{OCF}(\Sigma, \omega_1 \vee \omega_2)$ with exactly two worlds $\omega_1, \omega_2 \in \Omega_\Sigma$ in its domain it holds that $\xi_{\Sigma,\omega_1 \vee \omega_2}(\kappa_1) = a \cdot \kappa_1$. In the second part, we show that for any signature Σ' and any OCF $\kappa_6 \in \mathcal{M}_{OCF}(\Sigma', \omega_7 \vee \omega_8)$ with exactly two worlds $\omega_7, \omega_8 \in \Omega_{\Sigma'}$ in its domain it holds that $\xi_{\Sigma',A}(\kappa_6) = a \cdot \kappa$ (for the same a that we found in the first part). Finally, in the third part, we show that for any signature Σ', any formula $A \in \mathcal{L}_{\Sigma'}$, and any

OCF $\kappa \in \mathcal{M}_{OCF}(\Sigma', A)$ it holds that $\xi_{\Sigma',A}(\kappa) = a \cdot \kappa$, still for the a we found in the first part.

Part 1: Show (5) for restricted OCFs with a domain of two worlds over a single Σ

Let Σ be a signature with $|\Sigma| \geqslant 3$; thus $|\Omega_\Sigma| \geqslant 8$. First, we consider OCFs over Σ with two worlds in their domain. We need to show that there is an a such that for any $\omega_1, \omega_2 \in \Omega_\Sigma$ and $\kappa_1 \in \mathcal{M}_{OCF}(\Sigma, \omega_1 \vee \omega_2)$ Equation (5) holds. At least one of the worlds has to have rank 0 in κ_1; w.l.o.g. assume κ_1 has the form $\{\omega_1 \mapsto 0, \omega_2 \mapsto b\}$ for some $b \in \mathbb{N}_0$. Let $\kappa_1' = \xi(\kappa_1)$.

Let $\omega_3, \omega_4 \in \Omega_\Sigma$ be worlds and $\kappa_2 : \{\omega_3, \omega_4\} \to \mathbb{N}_0 \in \mathcal{M}_{OCF}(\Sigma, \omega_3 \vee \omega_4)$ with $\kappa_2(\omega_3) = 0, \kappa_2(\omega_4) = 1$. Let $\kappa_2' = \xi(\kappa_2)$. Let $a = \kappa_2'(\omega_4) - \kappa_2'(\omega_3)$. Because of (Ord), we have $a > 0$. Note that a is independent of the considered ranking function κ_1.

To show that $\kappa_1' = a \cdot \kappa_1$ we distinguish three cases.

Case 1.1: $b = 0$ For $b = 0$, (Ord) requires that κ_1' has the form $\{\omega_1 \mapsto 0, \omega_2 \mapsto 0\}$. Therefore, it holds that $\kappa_1' = a \cdot \kappa_1$.

Case 1.2: $b \geqslant 1$ and $\{\omega_1, \omega_2\} \cap \{\omega_3, \omega_4\} = \emptyset$ We show that $\kappa_1'(\omega_2) = a \cdot b$ by induction over b.

Base Case: Let $b = 1$. Consider the ranking function $\kappa_3 : \{\omega_1, \omega_2, \omega_3, \omega_4\} \to \mathbb{N}_0$ with $\kappa_3(\omega_1) = \kappa_3(\omega_3) = 0$ and $\kappa_3(\omega_2) = \kappa_3(\omega_4) = 1$. Let $\kappa_3' = \xi(\kappa_3)$. (Ord) requires that $\kappa_3'(\omega_1) = \kappa_3'(\omega_3) = 0$ and $\kappa_3'(\omega_2) = \kappa_3'(\omega_4)$. (Cond) requires that $\kappa_3'(\omega_4) = a$ because $\kappa_3|(\omega_3 \vee \omega_4) = \kappa_2$. Therefore, $\kappa_3'(\omega_2) = a = a \cdot b$. With (Cond) it follows that $\kappa_1'(\omega_2) = a \cdot b$.

Induction Step: Let $b = n + 1$. Let $\omega_5 \in \Omega_\Sigma \setminus \{\omega_1, \omega_2, \omega_3, \omega_4\}$ be an additional world. Consider the OCF $\kappa_4 : \{\omega_1, \omega_2, \omega_5\} \to \mathbb{N}_0$ with $\kappa_4(\omega_1) = 0, \kappa_4(\omega_5) = n$, and $\kappa_4(\omega_2) = n + 1$. Let $\kappa_4' = \xi(\kappa_4)$. The induction hypothesis in combination with (Cond) requires that $\kappa_4'(\omega_5) = a \cdot n$. Considering (Cond) and κ_2 as in the base case requires $\kappa_4'(\omega_2) - \kappa_4'(\omega_5) = a$. Hence, $\kappa_4'(\omega_2) = a \cdot (n + 1)$. With (Cond) it follows that $\kappa_1'(\omega_2) = a \cdot b$ and therefore $\kappa_1' = a \cdot \kappa_1$.

Case 1.3: $b \geqslant 1$ and $\{\omega_1, \omega_2\} \cap \{\omega_3, \omega_4\} \neq \emptyset$ Let $\omega_5, \omega_6 \in \Omega_\Sigma \setminus \{\omega_1, \omega_2, \omega_3, \omega_4\}$ be two additional worlds. Consider the OCF $\kappa_5 : \{\omega_1, \omega_2, \omega_5, \omega_6\} \to \mathbb{N}_0$ with $\kappa_5(\omega_1) = \kappa_5(\omega_3) = 0$ and $\kappa_5(\omega_2) = \kappa_5(\omega_4) = b$. Using *Case 1.2* we know that Equation (5) holds for $\kappa_5|(\omega_5 \vee \omega_6)$. By using (Cond) and (Ord) as in the Base Case of the induction proof in *Case 1.2*, we get $\kappa_1' = a \cdot \kappa_1$.

Part 2: Show (5) for restricted OCFs with a domain of two worlds over any sig. Σ'

We need to show that for any signature Σ', worlds $\omega_7, \omega_8 \in \Omega_{\Sigma'}$ and ranking function $\kappa_6 \in \mathcal{M}_{OCF}(\Sigma, \omega_7 \vee \omega_8)$, Equation (5) holds. At least one of the worlds has to have rank 0 in κ_6; w.l.o.g. assume κ_6 has the form $\{\omega_7 \mapsto 0, \omega_8 \mapsto c\}$ for some $c \in \mathbb{N}_0$. Let $\kappa_6' = \xi(\kappa_6)$. Because of (Ord), κ_6' has the form $\{\omega_7 \mapsto 0, \omega_8 \mapsto d\}$ for some $d \in \mathbb{N}_0$. It is left to show that $d = a \cdot c$.

We distinguish two cases:

Case 2.1: $\Sigma \not\subseteq \Sigma'$ Let $\omega_7^+, \omega_8^+ \in \Omega_{\Sigma \cup \Sigma'}$ be variable assignments such that the variables in Σ' are assigned to the same values in ω_7 and ω_7^+, the variables in Σ' are assigned to the same values in ω_8 and ω_8^+, and the variables in $\Sigma \setminus \Sigma'$ have the different values in ω_7^+ and ω_8^+. Let $\kappa_7 : \{\omega_7^+, \omega_8^+\} \to \mathbb{N}_0 \in \mathcal{M}_{OCF}(\Sigma \cup \Sigma', \omega_7^+ \vee \omega_8^+)$ with $\kappa_7(\omega_7^+) = 0$ and $\kappa_7(\omega_8^+) = c$ be a restricted ranking function over $\Sigma \cup \Sigma'$. Postulate (Marg) requires that $\xi(\kappa_7)_{|\Sigma'} = \xi(\kappa_{7|\Sigma'})$. Additionally, we have $\kappa_{7|\Sigma'} = \kappa_6$. Hence, we have that

$$\xi(\kappa_7)(\omega_8^+) = \xi(\kappa_7)_{|\Sigma'}(\omega_8) = \xi(\kappa_6)(\omega_8) = \kappa_6'(\omega_8) = d. \tag{6}$$

Let $\omega_7^\Sigma, \omega_8^\Sigma \in \Omega_\Sigma$ be the worlds over Σ such that ω_7^Σ assigns the variables in Σ to the same values as ω_7^+ and ω_8^Σ assigns the variables in Σ to the same values as ω_8^+. Because we chose different variable assignments over $\Sigma \setminus \Sigma'$ in ω_7^+, ω_8^+ we have that $\omega_7^\Sigma \neq \omega_8^\Sigma$. Marginalization of κ_7 to Σ yields the ranking function $\kappa_8 : \{\omega_7^\Sigma, \omega_8^\Sigma\} \to \mathbb{N}_0$ with $\kappa_8(\omega_7^\Sigma) = 0$ and $\kappa_8(\omega_8^\Sigma) = c$. With Part 1 of the proof it follows that $\xi(\kappa_8)(\omega_7^\Sigma) = 0$ and $\xi(\kappa_8)(\omega_8^\Sigma) = a \cdot c$. Analogously to (6) we have that $\xi(\kappa_7)(\omega_8^+) = \xi(\kappa_7)_{|\Sigma}(\omega_8^\Sigma) = \xi(\kappa_8)(\omega_8^\Sigma) = a \cdot c$. Together, we have $d = \xi(\kappa_6)(\omega_8^+) = a \cdot c$.

Note that *Case 2.1* also applies to signatures Σ' with one element.

Case 2.2: $\Sigma \subseteq \Sigma'$ Consider an additional signature Σ'' with $|\Sigma''| = |\Sigma|$ and $\Sigma'' \cap \Sigma' = \emptyset$. Equation (5) holds for OCFs with two worlds in their domain over Σ'' as *Case 2.1* applies. We can show $d = a \cdot c$ analogously to *Case 2.1* using Σ'' instead of Σ.

Part 3: Show Equation (5) for (restricted) OCFs over any number of worlds

We need to show that for any signature Σ', formula $A \in \mathcal{L}_{\Sigma'}$ and ranking function $\kappa \in \mathcal{M}_{OCF}(\Sigma, A)$, Equation (5) holds. Let $\omega_0 \in \kappa^{-1}(0)$ and $\kappa' = \xi(\kappa)$. Because of (Ord) and because the minimal rank of an OCF is 0, we have that $\kappa'(\omega_0) = 0$. Let $\omega \in dom(\kappa)$ be any world in the domain of κ. Let $\kappa_9 = \kappa|(\omega_0 \vee \omega)$ and let $\kappa_9' = \xi(\kappa_9)$. From Part 2 it follows that $\kappa_9' = a \cdot \kappa_9$. Using (Cond) we have $\kappa'(\omega) = \kappa'|(\omega_0 \vee \omega)(\omega) = \kappa_9'(\omega) = a \cdot \kappa_9(\omega) = a \cdot \kappa(\omega)$. □

As a direct implication of this, for epistemic state mappings from OCFs to OCFs (Ord), (Cond), and (Marg) imply (SynSplit).

Proposition 6. *For epistemic state mappings* $\xi : OCF \rightsquigarrow OCF$ *the conjunction of (Ord), (Cond), and (Marg) implies (SynSplit).*

In Proposition 6 postulate (Ord) can be replaced by the equivalent (IE), i.e., for $\xi : OCF \rightsquigarrow OCF$, (IE), (Cond), and (Marg) also imply (SynSplit).

5.3 Mapping Ranking Functions to Total Preorders

In this section, we investigate epistemic state mappings from ranking functions to total preorders on worlds. As for mappings from total preorders to total preorders, there is a unique mapping satisfying (Ord) and the equivalent (IE).

Proposition 7 (τ^*). *There is a unique epistemic state mapping $\tau^* : OCF \rightsquigarrow TPO$ fulfilling (Ord). This mapping is $\tau^* : OCF \rightsquigarrow TPO$ given by $\kappa \mapsto \preceq_\kappa$.*

Proof. Let κ be any ranking function. (Ord) states that $\preceq = \xi(\kappa)$ induces the same ranking function as κ. As the total preorder induced by a total preorder is the total preorder itself, the only epistemic state mapping from ranking functions to total preorders fulfilling (Ord) is

$$\tau^* : OCF \rightsquigarrow TPO, \quad \kappa \mapsto \preceq_\kappa.$$

□

In the following, we will investigate the properties of the epistemic state mapping τ^*. Obviously, τ^* is surjective: For a given total preorder $\preceq$ it is easy to construct a ranking function κ such that $\preceq = \tau^*(\kappa)$. But τ^* is not injective as there are more ranking functions than total preorders for any given (non-empty) signature (cf. [3]).

The epistemic state mapping τ^* preserves syntax splittings of the ranking function.

Proposition 8. *τ^* fulfils (SynSplit).*

Proof. Let $\kappa = \kappa_1 \oplus \cdots \oplus \kappa_n$ be an OCF over Σ with a syntax splitting $\{\Sigma_1, \ldots, \Sigma_n\}$. Let $\preceq = \tau^*(\kappa)$. Let $i \in \{1, \ldots, n\}$ and $\omega_1, \omega_2 \in \Omega_\Sigma$ with $\omega_1^{\neq i} = \omega_2^{\neq i}$ and $\omega_1 \preceq \omega_2$. Because τ^* fulfils (Ord), we have $\kappa(\omega_1) \leqslant \kappa(\omega_2)$. The syntax splitting on κ and $\omega_1^{\neq i} = \omega_2^{\neq i}$ implies $\kappa_i(\omega_1^i) \leqslant \kappa_i(\omega_2^i)$. This and the syntax splitting on κ implies $\omega_1^i \preceq_{|\Sigma_i} \omega_2^i$.

Thus, $\{\Sigma_1, \ldots, \Sigma_n\}$ is a syntax splitting for $\preceq$. □

As a direct implication of Proposition 8, τ^* fulfils (SynSplit[b]). However, τ^* may introduce *new* syntax splittings as the ranking function κ_2 in the following example shows.

Example 4. Let $\Sigma = \{a, b\}$ and κ_1, κ_2 be ranking functions over Σ such that:

$$\kappa_1(ab) = 0 \qquad \kappa_1(a\bar{b}) = 1 \qquad \kappa_1(\bar{a}b) = 1 \qquad \kappa_1(\bar{a}\bar{b}) = 2$$
$$\kappa_2(ab) = 0 \qquad \kappa_2(a\bar{b}) = 1 \qquad \kappa_2(\bar{a}b) = 1 \qquad \kappa_2(\bar{a}\bar{b}) = 3$$

κ_1 has the syntax splitting $\{\{a\}, \{b\}\}$, while κ_2 does not have this syntax splitting. Both ranking functions are mapped to the total preorder $ab \prec a\bar{b}, \bar{a}b \prec \bar{a}\bar{b}$ by τ^* which has the syntax splitting $\{\{a\}, \{b\}\}$.

The function τ^* behaves nicely with respect to marginalization and conditionalization.

Proposition 9. τ^* fulfils (Marg).

Proof. Let $\kappa \in \mathcal{M}_{OCF}(\Sigma, A)$ be an OCF and $\Sigma_1 \subseteq \Sigma$. Let $\preceq_1 = \tau^*(\kappa_{|\Sigma_1})$ and $\preceq_2 = \tau^*(\kappa)$. Let $\omega_a, \omega_b \in \Omega_{\Sigma_1}$.

$$\omega_a \preceq_1 \omega_b$$
$$\Leftrightarrow \kappa_{|\Sigma_1}(\omega_a) \leqslant \kappa_{|\Sigma_1}(\omega_b)$$
$$\Leftrightarrow \min(\{\kappa(\omega') \mid \omega' \in \Omega_\Sigma, \omega'^1 = \omega_1\}, \leqslant) \leqslant \min(\{\kappa(\omega') \mid \omega' \in \Omega_\Sigma, \omega'^1 = \omega_2\}, \leqslant)$$
$$\Leftrightarrow \min(\{\omega' \mid \omega' \in \Omega_\Sigma, \omega'^1 = \omega_1\}, \preceq_2) \preceq_2 \min(\{\omega' \mid \omega' \in \Omega_\Sigma, \omega'^1 = \omega_2\}, \preceq_2)$$
$$\Leftrightarrow \omega_a \preceq_{2|\Sigma_1} \omega_b$$

□

Proposition 10. τ^* fulfils (Cond).

Proof. Let $\kappa \in \mathcal{M}_{OCF}(\Sigma, A)$ be a ranking function and $F \in \mathcal{L}_\Sigma$ such that $Mod_\Sigma(A) \cap Mod_\Sigma(F) \neq \emptyset$. Let $\preceq_1 = \tau^*(\kappa|F)$ and $\preceq_2 = \tau^*(\kappa)$. Let $\omega_1, \omega_2 \in Mod(A \wedge F)$.

$$\omega_1 \preceq_1 \omega_2$$
$$\Leftrightarrow \kappa_{|F}(\omega_1) \leqslant \kappa_{|F}(\omega_2)$$
$$\Leftrightarrow \kappa(\omega_1) \leqslant \kappa(\omega_2)$$
$$\Leftrightarrow \omega_1 \preceq_2 \omega_2$$
$$\Leftrightarrow \omega_1 \preceq_2|F \omega_2$$

□

Similar to Propositions 5 and 6 we can conclude that for epistemic state mappings from OCFs to TPOs, (Ord) (or the equivalent (IE)) implies (SynSplit), (Marg), and (Cond).

Proposition 11. *Let $\xi : OCF \rightsquigarrow TPO$ be an empistemic state mapping from OCFs to TPOs. If ξ fulfils (Ord) or (IE) then ξ fulfils (SynSplit), (Cond), and (Marg).*

If we require only (wOrd$^\Rightarrow$) to hold for an epistemic state mapping, this mapping may split layers in $<_\kappa$ thereby adding new conditional beliefs. The other way round, an epistemic state mapping fulfilling only (wOrd$^\Leftarrow$) is allowed to merge layers of $<_\kappa$ and thereby to remove conditional beliefs.

5.4 Mapping Total Preorders to Ranking Functions

Now we want to consider epistemic state mappings that map a total preorder to a ranking function. Since the functions in τ^* are not bijective, we cannot simply reverse them. On the contrary, there is more than one epistemic state mapping $\rho : TPO \rightsquigarrow OCF$ that fulfils (Ord). That is not surprising as a ranking function contains more information than a total preorder over the same domain. The additional information is the absolute distance between worlds. The functions in ρ need to fill in this missing information and there is some freedom to do this. One of the epistemic state mappings satisfying (Ord) is the following mapping ρ^*.

Definition 14. *We define the epistemic state mapping $\rho^* : TPO \rightsquigarrow OCF$ as follows. For $\preceq \in \mathcal{M}_{TPO}(\Sigma, A)$ let*

$$L_0^{\preceq} = \min(dom(\preceq), \preceq) \text{ and}$$
$$L_k^{\preceq} = \min(dom(\preceq) \setminus (L_0^{\preceq} \cup \cdots \cup L_{k-1}^{\preceq}), \preceq) \text{ for } k > 0.$$

We define $\rho^*(\preceq) = \kappa_{\preceq}$ where for every $\omega \in dom(\preceq)$ we choose $\kappa_{\preceq}(\omega) = k$ such that $\omega \in L_k^{\preceq}$.

Note that the construction of $\kappa_{\preceq} = \rho^*(\preceq)$ in Definition 14 is well-defined, as the sets $L_i^{\preceq}$ and $L_j^{\preceq}$ are disjoint for $i \neq j$; and for every $\omega \in dom(\preceq)$ there is a k with $\omega \in L_k^{\preceq}$. Every set $L_k^{\preceq}$ corresponds to the k-th layer of $\preceq$.

Example 5. *The epistemic state mapping ρ^* maps the total preorder $ab \prec \overline{a}b, a\overline{b} \prec \overline{a}\overline{b}$ over signature $\Sigma = \{a, b\}$ to $\kappa_{\preceq} : \{ab \mapsto 0; \overline{a}b \mapsto 1; a\overline{b} \mapsto 1; \overline{a}\overline{b} \mapsto 2\}$.*

Proposition 12. *The epistemic state mapping ρ^* fulfils (Ord).*

The epistemic state mapping ρ^* from Definition 14 is only one of many possible epistemic state mappings fulfilling (Ord). An obvious wish would be to to use this set of possibilities to choose a transformation that fulfils additional postulates such as (SynSplit). However, ρ^* does not fulfil (SynSplit), and more generally there is no epistemic state mapping ρ that fulfils both (Ord) and (SynSplit).

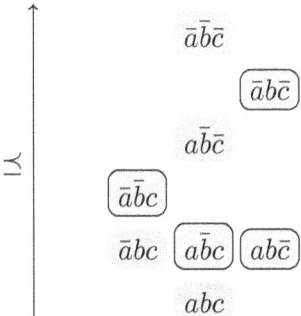

Figure 3: Total preorder $\preceq$ on $\Sigma = \{a, b, c\}$ with syntax splitting $\{\{a\}, \{b\}, \{c\}\}$. There is no ranking function with that syntax splitting that induces $\preceq$.

Proposition 13. *There is no epistemic state mapping $\rho : TPO \rightsquigarrow OCF$ fulfilling both (Ord) and (SynSplit).*

Proof. Let $\Sigma = \{a, b, c\}$ and $\preceq$ be the total preorder over Ω_Σ displayed in Figure 3. This TPO has the syntax splitting $\{\{a\}, \{b\}, \{c\}\}$. Consider the highlighted (red and circled) worlds in Figure 3. Let $\kappa = \rho(\preceq)$. Assume that (Ord) holds, i.e., the worlds $a\bar{b}c$ and $ab\bar{c}$ have the same rank in κ and the world $\bar{a}\bar{b}c$ has a lower rank than $\bar{a}b\bar{c}$ in κ. If κ had the syntax splitting $\{\{a\}, \{b\}, \{c\}\}$ it would also have the syntax splitting $\{\{a\}, \{b, c\}\}$ and therefore, we would have $\kappa(a\bar{b}c) - \kappa(ab\bar{c}) = \kappa(\bar{a}\bar{b}c) - \kappa(\bar{a}b\bar{c})$ leading to a contradiction. We conclude that there is no ranking function $\kappa = \rho(\preceq)$ such that both (Ord) holds and κ has the syntax splitting $\{\{a\}, \{b\}, \{c\}\}$. $\square$

The incompatibility observed in Proposition 13 persists even if we consider the weaker (SynSplit^b) instead of (SynSplit) and the weaker (wOrd$^\Rightarrow$) instead of (Ord).

Proposition 14. *There is no epistemic state mapping $\rho : TPO \rightsquigarrow OCF$ that fulfils both (wOrd$^\Rightarrow$) and (SynSplit^b).*

Proof. Let $\Sigma = \{a, b, c, d\}$ be a signature and $\preceq$ be the total preorder over Ω_Σ displayed in Figure 4. This TPO has the syntax splitting $\{\{a, b\}, \{c, d\}\}$. Towards a contradiction, assume there is a ranking function κ with syntax splitting $\{\{a, b\}, \{c, d\}\}$ such that $\omega_1 \prec \omega_2$ implies $\kappa(\omega_1) < \kappa(\omega_2)$. Then there are ranking functions $\kappa_1 : \Omega_{\{a,b\}} \to \mathbb{N}_0$ and $\kappa_2 : \Omega_{\{c,d\}} \to \mathbb{N}_0$ such that $\kappa = \kappa_1 \oplus \kappa_2$. Let

$$\kappa_1(ab) = 0 \qquad \kappa_1(a\bar{b}) = i \qquad \kappa_1(\bar{a}b) = j \qquad \kappa_1(\bar{a}\bar{b}) = k$$
$$\kappa_2(cd) = 0 \qquad \kappa_2(c\bar{d}) = l \qquad \kappa_2(\bar{c}d) = m \qquad \kappa_2(\bar{c}\bar{d}) = n.$$

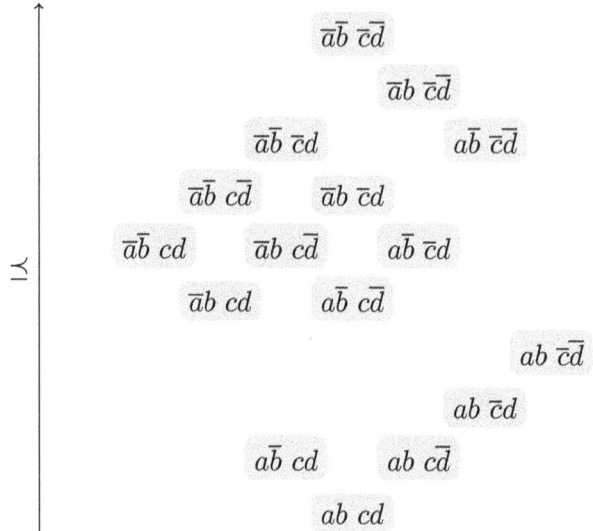

Figure 4: Total preorder $\preceq$ on $\Sigma = \{a, b, c, d\}$ with syntax splitting $\{a,b\} \mathbin{\dot\cup} \{c,d\}$. There is no ranking function with that syntax splitting that induces a superset of $\preceq$.

As $\omega_1 \prec \omega_2$ implies $\kappa(\omega_1) < \kappa(\omega_2)$ for every $\omega_1, \omega_2 \in \Omega_\Sigma$ we have that

$$m + j = \kappa_1(\overline{a}b) + \kappa_2(\overline{c}d) = \kappa(\overline{a}b\overline{c}d) < \kappa(a\overline{b}\overline{c}\overline{d}) = \kappa_1(a\overline{b}) + \kappa_2(\overline{c}\overline{d}) = i + n.$$

Analogously, we get $j > n$ from $\kappa(\overline{a}bcd) > \kappa(ab\overline{c}\overline{d})$ and $m > i$ from $\kappa(ab\overline{c}d) > \kappa(\overline{a}\overline{b}cd)$. The combination of these inequations is a contradiction. The assumed ranking function κ cannot exist. $\square$

However, the combination of (wOrd$^{\Leftarrow}$) and (SynSplit) can be fulfilled simultaneously, as we will see later in this section (Proposition 19).

Any epistemic state mapping ρ from total preorders to ranking functions satisfying (Ord) is compatible with τ^* (see Proposition 7) with respect to marginalization in the following sense.

Proposition 15. *Let* $\rho : TPO \rightsquigarrow OCF$ *be an epistemic state mapping that fulfils* (Ord). *For every total preorder* $\preceq \,\in \mathcal{M}_{TPO}(\Sigma, A)$ *and* $\Sigma' \subseteq \Sigma$ *it holds that*

$$\tau^*(\rho(\preceq)_{|\Sigma'}) = \preceq_{|\Sigma'}.$$

Proof. Let ρ satisfy (Ord). Let $\preceq$ be a TPO over $Mod_\Sigma(A)$ and $\kappa = \rho(\preceq)$. Let

Epistemic State Mappings

$\Sigma' \subseteq \Sigma$ and $\preceq' = \tau^*(\kappa_{|\Sigma'})$. Let $\omega_1, \omega_2 \in dom(\preceq')$.

$$\omega_1 \preceq' \omega_2$$
$$\Leftrightarrow \quad \kappa_{|\Sigma'}(\omega_1) \leqslant \kappa_{|\Sigma'}(\omega_2)$$
$$\Leftrightarrow \quad \min(\{\kappa(\omega) \mid \omega \in \Omega_\Sigma, \omega_1 \models \omega\}, \leqslant) \leqslant \min(\{\kappa(\omega) \mid \omega \in \Omega_\Sigma, \omega_2 \models \omega\}, \leqslant)$$
$$\Leftrightarrow \quad \min(\{\omega \mid \omega \in \Omega_\Sigma, \omega_1 \models \omega\}, \preceq) \preceq \min(\{\omega \mid \omega \in \Omega_\Sigma, \omega_2 \models \omega\}, \preceq)$$
$$\Leftrightarrow \quad \omega_1 \preceq_{|\Sigma'} \omega_2$$

□

Thus, an epistemic state mapping ρ from OCFs to TPOs fulfilling (Ord) commutes with τ^* and marginalization (cf. Figure 5a). A corresponding property of τ^* also holds for conditionalization (cf. Figure 5b).

Proposition 16. *Let $\rho : TPO \rightsquigarrow OCF$ be an epistemic state mapping fulfilling (Ord). For every total preorder $\preceq \in \mathcal{M}_{TPO}(\Sigma, A)$ and $F \in \mathcal{L}_\Sigma$ it holds that*

$$\tau^*(\rho(\preceq)|F) = \preceq|F.$$

Proof. Let ρ satisfy (Ord). Let $\preceq$ be a TPO over $Mod_\Sigma(A)$ and $\kappa = \rho(\preceq)$. Let $F \in \mathcal{L}_\Sigma$ and $\preceq' = \tau^*(\kappa|F)$. Let $\omega_1, \omega_2 \in Mod_\Sigma(A \wedge F)$.

$$\omega_1 \preceq' \omega_2$$
$$\Leftrightarrow \quad \kappa|F(\omega_1) \leqslant \kappa|F(\omega_2)$$
$$\Leftrightarrow \quad \kappa(\omega_1) \leqslant \omega_2$$
$$\Leftrightarrow \quad \omega_1 \preceq \omega_2$$
$$\Leftrightarrow \quad \omega_1 \preceq|F \omega_2$$

□

It would be useful, if a transformation from a total preorder to a ranking function preserved marginalization and conditionalization in the way τ^* does for transformations from OCFs to TPOs (see Propositions 9 and 10). However, Postulate (Cond) cannot be fulfilled simultaneously with (Ord). Moreover, (Cond) is even incompatible with the weaker Postulate (wOrd$^\Rightarrow$).

Proposition 17. *There is no epistemic state mapping $\rho : TPO \rightsquigarrow OCF$ that fulfils (Cond) and (wOrd$^\Rightarrow$).*

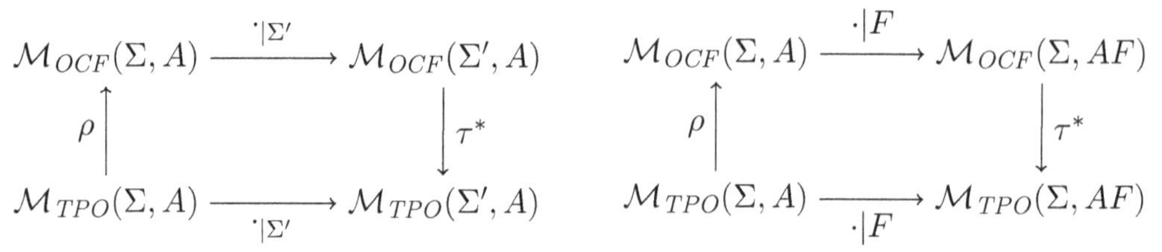

(a) Illustration of Proposition 15. (b) Illustrating of Proposition 16.

Figure 5: Commutating diagrams illustrating Propositions 15 and 16. Precondition of both propositions is that $\rho : TPO \rightsquigarrow OCF$ fulfils (Ord).

Figure 6: Total preorders $\preceq_1$ and $\preceq_2$ on $\Sigma = \{a,b\}$ which show that (Cond) is incompatible with (wOrd$^\Rightarrow$) for epistemic state mappings from TPOs to OCFs.

Proof. Let $\Sigma = \{a,b\}$ be a signature and $\preceq_1, \preceq_2$ be the total preorders over Ω_Σ displayed in Figure 6. We have $\preceq_1|a = \preceq_2|a$ and $\preceq_1|b = \preceq_2|b$. Let $\kappa_1 = \rho(\preceq_1)$ and $\kappa_2 = \rho(\preceq_2)$. If (wOrd$^\Rightarrow$) and (Cond) were fulfilled it would imply $\kappa_1(\bar{a}b) = (\kappa_1|b)(\bar{a}) = (\kappa_2|b)(\bar{a}) = \kappa_2(\bar{a}b)$ and $\kappa_1(a\bar{b}) = (\kappa_1|a)(\bar{b}) = (\kappa_2|a)(\bar{b}) = \kappa_2(a\bar{b})$. This contradicts (wOrd$^\Rightarrow$) as (wOrd$^\Rightarrow$) requires $\kappa_1(\bar{a}b) > \kappa_1(a\bar{b})$ and $\kappa_2(\bar{a}b) < \kappa_2(a\bar{b})$. $\square$

Also, postulate (Marg) is in general unfulfillable in combination with (wOrd$^\Rightarrow$) for epistemic state mappings from TPOs to OCFs.

Proposition 18. *There is no epistemic state mapping $\rho : TPO \rightsquigarrow OCF$ that fulfils both (wOrd$^\Rightarrow$) and (Marg).*

Proof. Let $\Sigma = \{a,b\}$ be a signature and $\preceq_1, \preceq_2$ be the total preorders over Ω_Σ displayed in Figure 7. Let $\Sigma_1 = \{a\}$ and $\Sigma_2 = \{b\}$. We have $\preceq_1|_{\Sigma_1} = \preceq_2|_{\Sigma_1}$ and $\preceq_1|_{\Sigma_2} = \preceq_2|_{\Sigma_2}$. Let $\kappa_1 = \rho(\preceq_1)$ and $\kappa_2 = \rho(\preceq_2)$. If (wOrd$^\Rightarrow$) and (Marg) were fulfilled it would imply $\kappa_1(\bar{a}b) = \kappa_{1|\Sigma_1}(\bar{a}b) = \kappa_{2|\Sigma_1}(\bar{a}b) = \kappa_2(\bar{a}b)$ and $\kappa_1(a\bar{b}) = \kappa_{1|\Sigma_2}(a\bar{b}) = \kappa_{2|\Sigma_2}(a\bar{b}) = \kappa_2(a\bar{b})$. This contradicts (wOrd$^\Rightarrow$) as (wOrd$^\Rightarrow$) requires $\kappa_1(\bar{a}b) > \kappa_1(a\bar{b})$ and $\kappa_2(\bar{a}b) < \kappa_2(a\bar{b})$. $\square$

Figure 7: Total preorders $\preceq_1$ and $\preceq_2$ on $\Sigma = \{a, b\}$ which show that (Marg) is incompatible with (wOrd$^\Rightarrow$) for epistemic state mappings from TPOs to OCFs.

Propositions 14, 17, and 18 all showed that (wOrd$^\Rightarrow$) resp. the equivalent (wIE$^\Rightarrow$) cannot be fulfilled in combination with (SynSplit), (SynSplit^b), (Cond), or (Marg). Postulate (wOrd$^\Leftarrow$) on the other hand can be fulfilled in combination with these other postulates.

Proposition 19. *There is an epistemic state mapping $\xi :$ TPO $\rightsquigarrow$ OCF fulfilling (wOrd$^\Leftarrow$), (SynSplit), (SynSplit^b), (Cond), and (Marg).*

Proof. The epistemic state mapping that maps every TPO to the uniform ranking function κ_{uni} over the respective domain fulfils all three postulates. $\square$

However, the following triviality result shows that there is only one epistemic state mapping fulfilling the combination of (wIE$^\Leftarrow$) and (Cond).

Proposition 20. *The only epistemic state mapping $\rho :$ TPO $\rightsquigarrow$ OCF that fulfils (wIE$^\Leftarrow$) and (Cond) maps every TPO to the trivial uniform ranking function κ_{uni}.*

Proof. Let ρ be an epistemic state mapping fulfilling (wIE$^\Leftarrow$) and (Cond). Let Σ be a signature and $\omega_1, \omega_2 \in \Omega_\Sigma$ with $\omega_1 \neq \omega_2$. Choose a third world $\omega_3 \in \Omega_\Sigma$ with $\omega_3 \notin \{\omega_1, \omega_2\}$ and consider the TPOs

$$\omega_3 \prec_1 \omega_2 \prec_1 \omega_1 \prec_1 \omega_4, \ldots, \omega_n$$
$$\omega_3 \prec_2 \omega_1 \prec_2 \omega_2 \prec_2 \omega_4, \ldots, \omega_n$$

with $\{\omega_4, \ldots, \omega_n\} = \Omega_\Sigma \setminus \{\omega_1, \omega_2, \omega_3\}$. Let $\kappa_1 = \rho(\preceq_1)$ and $\kappa_2 = \rho(\preceq_2)$. The postulate (wIE$^\Leftarrow$) requires that

$$\kappa_1(\omega_2) \leq \kappa_1(\omega_1) \quad \text{and} \quad \kappa_2(\omega_1) \leq \kappa_2(\omega_2). \qquad (*)$$

Let $A = \omega_3 \vee \omega_1$ and $B = \omega_3 \vee \omega_2$. Conditionalization yields $\preceq'_A = \preceq_1|A = \preceq_2|A$ and $\preceq'_B = \preceq_1|B = \preceq_2|B$. Postulate (Cond) requires $\kappa_1|A = \rho(\preceq'_A) = \kappa_2|A$ and

$\kappa_1|B = \rho(\preceq'_B) = \kappa_2|B$. This implies $\kappa_1(\omega_1) = \kappa_2(\omega_1)$ and $\kappa_1(\omega_2) = \kappa_2(\omega_2)$. With (*) it follows that $\kappa_1(\omega_2) \leqslant \kappa_1(\omega_1) = \kappa_2(\omega_1) \leqslant \kappa_2(\omega_2) = \kappa_2(\omega_2)$. Therefore we can replace the $\leqslant$ in this chain of (in-)equations by $=$. Let $C = \omega_1 \vee \omega_2$. We can see that both $\preceq_1|C = \{\omega_1 \prec \omega_2\}$ and $\preceq_2|C = \{\omega_2 \prec \omega_1\}$ are mapped to the uniform ranking function κ_{uni} due to (Cond). With (wIE$^\Leftarrow$) follows that ω_1, ω_2 have to be mapped to the same rank in every ranking function.

Since we can choose any two worlds as ω_1, ω_2 in this argumentation, all worlds must have the same rank in the resulting ranking function. Therefore, any TPO is mapped to the uniform ranking function κ_{uni} by ρ. □

Considering the results in this section, other postulates beside (SynSplit), (Cond), and (Marg) may be necessary to guide epistemic state mappings from TPOs to OCFs.

6 Mapping Coherent Total Preorders to Ranking Functions

In Section 5.4 we saw that there is no epistemic state mapping from TPOs to OCFs that fulfils (Ord) and one of the postulates (Cond), (Marg), or (SynSplit^b) simultaneously.

One reason for the incompatibility results in the previous section is that the number of layers between two worlds can change during marginalization and conditionalization of TPOs. In this section we will consider weaker versions of the postulates (Cond), (Marg), (SynSplit) and (SynSplit^b) that can be fulfilled simultaneously by an epistemic state mapping from TPOs to OCFs. For this purpose, we strengthen the prerequisite of the postulates by employing the notion of coherence.

Coherence describes that a set of worlds has no "gaps" in a total preorder. A set of worlds M is coherent with respect to a total preorder $\preceq$ if there are no two worlds $\omega_1, \omega_3 \in M$ such that ω_3 follows ω_1 directly in M but not in the domain of $\preceq$; i.e., if there is no world $\omega'_2 \in M$ with $\omega_1 \prec \omega'_2 \prec \omega_3$ then there cannot be an $\omega_2 \in dom(\preceq)$ with $\omega_1 \prec \omega_2 \prec \omega_3$.

Definition 15 (coherence for TPOs [23]). *Let $\preceq \in \mathcal{M}_{TPO}(\Sigma)$ be a total preorder. We call $M \subseteq \Omega_\Sigma$ coherent with respect to $\preceq$ if for every $\omega_1, \omega_2, \omega_3 \in \Omega_\Sigma$ with $\omega_1, \omega_3 \in M$ and $\omega_1 \prec \omega_2 \prec \omega_3$ there is a world $\omega'_2 \in M$ such that $\omega'_2 \approx \omega_2$.*

A formula $F \in \mathcal{L}_\Sigma$ is coherent with respect to $\preceq$ if $Mod_\Sigma(F)$ is coherent with respect to $\preceq$.

The notation of coherence can be extended to restricted TPOs.

Definition 16 (coherence for restricted TPOs). *Let $\preceq\, \in \mathcal{M}_{TPO}(\Sigma, A)$ be a restricted TPO. We call $M \subseteq dom(\preceq)$ coherent with respect to $\preceq$ if for every $\omega_1, \omega_2, \omega_3 \in dom(\preceq)$ with $\omega_1, \omega_3 \in M$ and $\omega_1 \prec \omega_2 \prec \omega_3$ there is a world $\omega_2' \in M$ such that $\omega_2' \approx \omega_2$.*

A formula $F \in \mathcal{L}_\Sigma$ is coherent with respect to $\preceq$ if $Mod_\Sigma(F \wedge A)$ is coherent with respect to $\preceq$.

Example 6. *The TPO $\preceq_1$ shown in Figure 7 is coherent with respect to a, but not coherent with respect to b. The TPO $\preceq_1$ in Figure 6 is coherent with respect to $\bar{b}$, but not coherent with respect to b.*

Using coherence, we can weaken the postulate (Cond) by restricting it to TPOs that are coherent with respect to the formula used for conditionalization.

Postulates. *Let $(\xi_{\Sigma,A})$ be an epistemic state mapping from TPOs to OCFs. Let Σ be a signature, $A \in \mathcal{L}_\Sigma$, and $\preceq\, \in \mathcal{M}_{TPO}(\Sigma, A)$.*

Let $\Sigma' \subseteq \Sigma$ with $\Sigma' \neq \emptyset$ and $A' = A_{|\Sigma'}$. Let $F \in \mathcal{L}_\Sigma$ with $Mod_\Sigma(F) \cap dom(\Psi) \neq \emptyset$.

(Cond$_{coh}$) *If $\preceq$ is coherent with respect to F, then $\xi_{\Sigma, A \wedge F}(\preceq|F) = \xi_{\Sigma, A}(\preceq)|F$.*

Obviously, (Cond$_{coh}$) is a weaker version of (Cond).

Proposition 21. *Postulate (Cond) implies (Cond$_{coh}$).*

Consider again the epistemic state mapping ρ^* that maps a TPO $\preceq$ to $\kappa_\preceq$ (see Definition 14). From the results in Section 5.3 we know that ρ^* cannot fulfil (SynSplit^b), (Marg), and (Cond) as it fulfils (Ord). However, it fulfils weaker notions of these postulates based on coherence. To prove that (Cond$_{coh}$) is fulfilled by ρ^*, we use the following definition of the *distance of two worlds* in a total preorder.

Definition 17 ($dist_\preceq(\omega, \omega')$). *Let $\preceq$ be a restricted total preorder on worlds and $\omega, \omega' \in dom(\preceq)$ such that $\omega \preceq \omega'$. The distance $dist_\preceq(\omega, \omega')$ between ω and ω' in $\preceq$ is the length of the longest chain of inequations between ω and ω' of the form*

$$\omega \prec \omega_1 \prec \omega_2 \prec \cdots \prec \omega_{n-1} \prec \omega'.$$

with $\omega_1, \ldots, \omega_{n-1} \in dom(\preceq)$. If $\omega \approx \omega'$ then $dist_\preceq(\omega, \omega') = 0$.

Note that if $\omega \prec \omega'$ and $\{\omega, \omega'\}$ is coherent with respect to $\preceq$, then $dist_\preceq(\omega, \omega') = 1$.

Lemma 2. *Let $\preceq$ be a (restricted) total preorder. For two worlds $\omega, \omega' \in dom(\preceq)$ it holds that*
$$dist_{\preceq}(\omega, \omega') = \kappa_{\preceq}(\omega) - \kappa_{\preceq}(\omega').$$

Lemma 3. *Let $\preceq$ be a (restricted) total preorder and $M \subseteq dom(\preceq)$ a set of worlds such that $\preceq$ is coherent with respect to M. For two worlds $\omega, \omega' \in M$ we have that $dist_{\preceq}(\omega, \omega') = dist_{\preceq|M}(\omega, \omega')$.*

Using the two lemmas above we show that ρ^* fulfils (Cond$_{coh}$). The following proposition generalizes a result of [23] to the setting of restricted TPOs and OCFs introduced here.

Proposition 22 (Cond$_{coh}$). *The epistemic state mapping ρ^* fulfils (Cond$_{coh}$).*

Proof. Let Σ be a signature, $A \in \mathcal{L}_\Sigma$, and $\preceq \in \mathcal{M}_{OCF}(\Sigma, A)$ such that $\preceq$ is coherent with respect to $F \in \mathcal{L}_\Sigma$. We need to show that $\rho^*(\preceq|F) = \rho^*(\preceq)|F$.

Let $\omega \in Mod_\Sigma(F \wedge A)$ and $\omega' \in \min(Mod_\Sigma(F \wedge A), \preceq)$. In the following, we will denote $\rho^*(\preceq)$ with $\kappa_\preceq$. Let $n = dist_\preceq(\omega, \omega')$ be the distance between ω and ω' in $\preceq$. Because $\preceq$ is coherent with respect to F, we have $dist_{\preceq|F}(\omega, \omega') = dist_\preceq(\omega, \omega') = n$ (see Lemma 3). Furthermore, $dist_{\preceq|F}(\omega, \omega') = \kappa_{\preceq|F}(\omega) - \kappa_{\preceq|F}(\omega')$ (see Lemma 2). Hence, $\kappa_\preceq(\omega) - \kappa_\preceq(\omega') = n = \kappa_{\preceq|F}(\omega) - \kappa_{\preceq|F}(\omega')$. Because ω' was chosen minimal in $Mod_\Sigma(F \wedge A)$, we have $\kappa_{\preceq|F}(\omega') = 0$. With this, we have

$$\kappa_\preceq|F(\omega) = \kappa_\preceq(\omega) - \kappa_\preceq(F) = \kappa_\preceq(\omega) - \kappa_\preceq(\omega') = \kappa_{\preceq|F}(\omega) - \kappa_{\preceq|F}(\omega') = \kappa_{\preceq|F}(\omega)$$

where the latter equation holds because $\kappa_{\preceq|F}(\omega') = 0$. Therefore, $\rho^*(\preceq|F) = \kappa_{\preceq|F} = \kappa_\preceq|F = \rho^*(\preceq)|F$. □

Similar to (Cond$_{coh}$), we can also weaken the postulate (Marg) for marginalization by employing the notion of coherence and requiring the set of models in $Mod_\Sigma(A)$ that are minimal models coinciding with a model from $Mod_{\Sigma'}(A')$ to be coherent.

Postulates. *Let $(\xi_{\Sigma, A})$ be an epistemic state mapping from TPOs to OCFs. Let Σ be a signature, $A \in \mathcal{L}_\Sigma$, and $\preceq \in \mathcal{M}_{TPO}(\Sigma, A)$.*

Let $\Sigma' \subseteq \Sigma$ with $\Sigma' \neq \emptyset$ and $A' = A_{|\Sigma'}$.

(Marg$_{coh}$) *If $\preceq$ is coherent with respect to*
$$M = \{\min(\{\omega \in dom(\preceq) \mid \omega' \models \omega\}, \preceq) \mid \omega' \in Mod_{\Sigma'}(A')\},$$
then $\xi_{\Sigma', A'}(\preceq_{|\Sigma'}) = \xi_{\Sigma, A}(\preceq)_{|\Sigma'}$.

The following observation is obvious.

Proposition 23. *Postulate (Marg) implies (Marg$_{coh}$).*

The set M in postulate (Marg$_{coh}$) contains the worlds determining the marginalized ordering $\preceq_{|\Sigma'}$. While ρ^* does not fulfil (Marg), we show that it fulfils (Marg$_{coh}$). For this proof we extend the notion of distance in a TPO to formulas.

Definition 18 ($dist_\preceq(A,B)$). *Let $\preceq$ be a restricted TPO over Ω_Σ and let A, B be formulas such that $Mod_\Sigma(A) \cap dom(\preceq) \neq \emptyset$ and $Mod_\Sigma(B) \cap dom(\preceq) \neq \emptyset$ and $A \preceq B$. The distance $dist_\preceq(A,B)$ between A and B in $\preceq$ is defined as the smallest distance between an element of $\min(Mod_\Sigma(A) \cap dom(\preceq), \preceq)$ and an element of $\min(Mod_\Sigma(B) \cap dom(\preceq), \preceq)$.*

Note that in Definition 18 for every $\omega \in \min(Mod_\Sigma(A) \cap dom(\preceq), \preceq)$ and every $\omega' \in \min(Mod_\Sigma(B) \cap dom(\preceq), \preceq)$ the distance $dist_\preceq(\omega, \omega')$ is the same. Hence, requiring the smallest distance between elements in this definition is not strictly necessary.

Lemma 4. *Let $\preceq$ be a restricted TPO over Ω_Σ and let A, B be formulas over $\Sigma' \subseteq \Sigma$ such that $Mod_\Sigma(A) \cap dom(\preceq) \neq \emptyset$ and $Mod_\Sigma(B) \cap dom(\preceq) \neq \emptyset$ and $A \preceq B$. If $\preceq$ is coherent with respect to $M = \{\min(\{\omega \in dom(\preceq) \mid \omega' \models \omega\}, \preceq) \mid \omega' \in Mod_{\Sigma'}(A')\}$, then $dist_\preceq(A,B) = dist_{\preceq_{|\Sigma'}}(A,B)$.*

Because the ordering $\preceq_{|\Sigma'}$ of the worlds over Σ' follows the ordering of the worlds M as in $\preceq$ and because the distances $dist_\preceq$ and $dist_\preceq$ are based on minimal worlds, marginalization does not change the distance of the formulas in Lemma 4. Using this lemma we can prove that ρ^* fulfils (Marg$_{coh}$).

Proposition 24 (Marg$_{coh}$). *ρ^* fulfils (Marg$_{coh}$).*

Proof. Let Σ be a signature, $A \in \mathcal{L}_\Sigma$ a formula, $\Sigma' \subseteq \Sigma$ a sub-signature, $A' = A_{|\Sigma'}$, and $\preceq \in \mathcal{M}_{OCF}(\Sigma, A)$ such that $\preceq$ is coherent with respect to $M = \{\min(\{\omega \in dom(\preceq) \mid \omega' \models \omega\}, \preceq) \mid \omega' \in Mod_{\Sigma'}(A')\}$. We need to show that $\rho^*(\preceq_{|\Sigma'}) = \rho^*(\preceq)_{|\Sigma'}$. Let $D' = dom(\preceq_{|\Sigma'}) = Mod_{\Sigma'}(A')$ and let $\omega \in D'$ and $\omega' \in \min(D', \preceq_{|\Sigma'})$. Then we have

$$dist_{\preceq_{|\Sigma'}}(\omega, \omega') = \kappa_{(\preceq_{|\Sigma'})}(\omega) - \kappa_{(\preceq_{|\Sigma'})}(\omega') = \kappa_{(\preceq_{|\Sigma'})}(\omega)$$

where the first equation holds because of Lemma 2 and the later equation holds because $\kappa_{(\preceq_{|\Sigma'})}(\omega') = 0$ as ω' is chosen minimally. Furthermore, $dist_{\preceq_{|\Sigma'}}(\omega, \omega') = dist_\preceq(\omega, \omega')$ because $\preceq$ is coherent with respect to M (see Lemma 4). Additionally,

$$dist_\preceq(\omega, \omega') = \kappa_\preceq(\omega) - \kappa_\preceq(\omega') \qquad \text{(see Lemma 2)}$$
$$= (\kappa_\preceq)_{|\Sigma'}(\omega) - (\kappa_\preceq)_{|\Sigma'}(\omega') = (\kappa_\preceq)_{|\Sigma'}(\omega) \qquad \text{(see Lemma 1)}$$

where the latter equation holds because $(\kappa_{\preceq})_{|\Sigma'}(\omega') = 0$ as ω' is chosen minimally. Hence, $\rho^*(\preceq_{|\Sigma'}) = \kappa_{\preceq_{|\Sigma'}} = dist_{\preceq_{|\Sigma'}}(\omega, \omega') = \kappa_{\preceq|\Sigma'} = \rho^*(\preceq)_{|\Sigma'}$. □

Analogously to (Cond$_{coh}$) and (Marg$_{coh}$) we present a weaker version of (SynSplit^b) based on coherence.

Postulates. *Let $(\xi_{\Sigma,A})$ be an epistemic state mapping from TPOs to OCFs. Let Σ be a signature, $A \in \mathcal{L}_\Sigma$, and $\preceq \in \mathcal{M}_{TPO}(\Sigma, A)$.*

(SynSplit$^b_{coh}$) *If $\{\Sigma_1, \Sigma_2\}$ is a syntax splitting for $\preceq$ and $\preceq$ is coherent with respect to*

$$\{\omega^1\omega^2 \mid \omega^1 \in dom(\preceq_{|\Sigma_1})\} \quad \text{for every } \omega^2 \in dom(\preceq_{|\Sigma_2}),$$

then $\{\Sigma_1, \Sigma_2\}$ is a syntax splitting for $\xi_{\Sigma,A}(\preceq)$.

Proposition 25. *Postulate (SynSplit^b) implies (SynSplit$^b_{coh}$).*

While ρ^* does not fulfil (SynSplit^b), it fulfills the weaker postulate (SynSplit$^b_{coh}$).

Lemma 5. *Let $\preceq$ be a total preorder and $\omega_1, \omega_2, \omega_3 \in dom(\preceq)$ be worlds such that $\omega_1 \preceq \omega_2 \preceq \omega_3$. Then $dist_{\preceq}(\omega_1, \omega_3) = dist_{\preceq}(\omega_1, \omega_2) + dist_{\preceq}(\omega_2, \omega_3)$.*

Proposition 26. *ρ^* fulfils (SynSplit$^b_{coh}$)*

Proof. Let Σ be a signature, $A \in \mathcal{L}_\Sigma$, and $\preceq \in \mathcal{M}_{TPO}(\Sigma, A)$ a TPO with syntax splitting $\{\Sigma_1, \Sigma_2\}$, such that $\preceq$ is coherent with respect to $\{\omega^1\omega^2 \mid \omega^1 \in dom(\preceq_{|\Sigma_1})\}$ for every $\omega^2 \in dom(\preceq_{|\Sigma_2})$. We need to show that $\{\Sigma_1, \Sigma_2\}$ is a syntax splitting for $\rho^*(\preceq)$.

Let $\omega \in dom(\preceq)$ be a world, ω^1 be the part of ω over Σ_1 and ω^2 be the part of ω over Σ_2. Let $D_{\Sigma_1} = dom(\preceq_{|\Sigma_1})$ and $D_{\Sigma_2} = dom(\preceq_{|\Sigma_2})$. Because of the syntax splitting, we have $dom(\preceq) = \{\omega^1\omega^2 \mid \omega_1 \in D_{\Sigma_1}, \omega_2 \in D_{\Sigma_2}\}$. Let $\omega_{min} \in \min(dom(\preceq), \preceq)$. We denote the part of ω_{min} over Σ_1 as ω^1_{min} and the part of ω_{min} over Σ_2 as ω^2_{min}. The rank of ω is $\kappa(\omega) = dist_{\preceq}(\omega_{min}, \omega) = dist_{\preceq}(\omega^1_{min}\omega^2_{min}, \omega^1_{min}\omega^2) + dist_{\preceq}(\omega^1_{min}\omega^2, \omega^1\omega^2)$ (see Lemma 5 and Lemma 2).

The syntax splitting ensures that the ordering of the set $\{\omega'^1\omega^2 \mid \omega'^1 \in D_{\Sigma_1}\}$ with respect to $\preceq$ is equivalent to the ordering of the set $\{\omega'^1\omega^2_{min} \mid \omega'^1 \in D_{\Sigma_1}\}$ with respect to $\preceq$. Hence, $\omega^1_{min}\omega^2$ is minimal in $Mod_\Sigma(\omega_2) = \{\omega'^1\omega^2 \mid \omega'^1 \in D_{\Sigma_1}\}$ and therefore $\kappa_{\preceq}(\omega_2) = \kappa_{\preceq}(\omega^1_{min}\omega^2)$. This yields

$$dist_{\preceq}(\omega^1_{min}\omega^2_{min}, \omega^1_{min}\omega^2) = \underbrace{\kappa_{\preceq}(\omega^1_{min}\omega^2)}_{=\kappa_{\preceq}(\omega_2)} - \underbrace{\kappa_{\preceq}(\omega^1_{min}\omega^2_{min})}_{=0} = \kappa_{\preceq}(\omega_2).$$

Because of the syntax splitting, the ordering of the set $\{\omega'^1\omega^2 \mid \omega'^1 \in D_{\Sigma_1}\}$ with respect to $\preceq$ is equivalent to the ordering of D_{Σ_1} with respect to $\preceq_{|\Sigma_1}$. Because of this and because $\{\omega'^1\omega^2 \mid \omega'^1 \in D_{\Sigma_1}\}$ is coherent with respect to $\preceq$ we have $dist_{\preceq}(\omega^1_{min}\omega^2, \omega^1\omega^2) = dist_{\preceq_{|\Sigma_1}}(\omega^1_{min}, \omega^1)$. Additionally, $dist_{\preceq_{|\Sigma_1}}(\omega^1_{min}, \omega^1) = \kappa_{\preceq_{|\Sigma_1}}(\omega_1)$ as ω^1_{min} is minimal with respect to $\preceq_{|\Sigma_2}$.

Combining these results yields $\kappa_{\preceq}(\omega) = \kappa_{\preceq_{|\Sigma_1}}(\omega_1) + \kappa_{\preceq}(\omega_2)$. Hence, $\kappa_{\preceq}$ has the syntax splitting $\{\Sigma_1, \Sigma_2\}$. □

Finally, we present a weaker version of (SynSplit) using coherence.

Postulates. *Let $(\xi_{\Sigma,A})$ be an epistemic state mapping from TPOs to OCFs. Let Σ be a signature, $A \in \mathcal{L}_\Sigma$, and $\preceq \in \mathcal{M}_{TPO}(\Sigma, A)$.*

(SynSplit$_{coh}$) *If $\{\Sigma_1, \ldots, \Sigma_n\}$ is a syntax splitting for $\preceq$ and $\preceq$ is coherent with respect to*

$$\{\omega^{\neq i}\omega^i \mid \omega^i \in dom(\preceq_{|\Sigma_i})\} \qquad \text{for every } \omega^{\neq i} \in dom(\preceq_{|\Sigma \setminus \Sigma_i})$$
$$\text{for every } i \in \{1, \ldots, n\},$$

then $\{\Sigma_1, \ldots, \Sigma_n\}$ is a syntax splitting for $\xi_{\Sigma,A}(\preceq)$.

Proposition 27. *Postulate (SynSplit) implies (SynSplit$_{coh}$).*

While ρ^* does not fulfil (SynSplit), it does satisfy (SynSplit$_{coh}$).

Proposition 28. *ρ^* fulfils (SynSplit$_{coh}$).*

Proof. Let Σ be a signature, $A \in \mathcal{L}_\Sigma$, and $\preceq \in \mathcal{M}_{TPO}(\Sigma, A)$ a TPO with syntax splitting $\{\Sigma_1, \ldots, \Sigma_n\}$, such that $\preceq$ is coherent with respect to $\{\omega^{\neq i}\omega^i \mid \omega^i \in dom(\preceq_{|\Sigma_i})\}$ for every $\omega^{\neq i} \in dom(\preceq_{|\Sigma \setminus \Sigma_i})$ for every $i \in \{1, \ldots, n\}$. We need to show that $\{\Sigma_1, \ldots, \Sigma_n\}$ is a syntax splitting for $\kappa = \rho^*(\preceq)$.

Let ω be any world in $dom(\preceq)$ and let ω_{min} be a world in $dom(\preceq)$ that is minimal with respect to $\preceq$. Let $\omega^{\leqslant 0}_{min} = \omega_{min}$, let $\omega^{\leqslant 1}_{min} = \omega^1 \omega^2_{min} \ldots \omega^n_{min}$, let $\omega^{\leqslant 2}_{min} = \omega^1 \omega^2 \omega^3_{min} \ldots \omega^n_{min}$, and so on until $\omega^{\leqslant n}_{min} = \omega$. Using Lemma 5 repeatedly we have

$$\kappa(\omega) = dist_{\preceq}(\omega_{min}, \omega)$$
$$= dist_{\preceq}(\omega^{\leqslant 0}_{min}, \omega^{\leqslant 1}_{min}) + dist_{\preceq}(\omega^{\leqslant 1}_{min}, \omega^{\leqslant 2}_{min}) + \cdots + dist_{\preceq}(\omega^{\leqslant n-1}_{min}, \omega^{\leqslant n}_{min}).$$

We have $dist_{\preceq}(\omega^{\leqslant k-1}_{min}, \omega^{\leqslant k}_{min}) = dist_{\preceq}(\omega^k_{min}, \omega^k) = \rho^*(\preceq_{|\Sigma_k})(\omega^k)$ for $k = 1, \ldots, n$ because $\{\omega^{\neq k}\omega'^k \mid \omega'^k \in dom(\preceq_{|\Sigma_k})\}$ is coherent. Hence, $\kappa(\omega) = dist_{\preceq}(\omega_{min}, \omega) = \rho^*(\preceq_{|\Sigma_1})(\omega^1) + \cdots + \rho^*(\preceq_{|\Sigma_n})(\omega^n)$ and therefore $\kappa = \rho^*(\preceq_{|\Sigma_1}) \oplus \cdots \oplus \rho^*(\preceq_{|\Sigma_n})$. This shows that $\{\Sigma_1, \ldots, \Sigma_n\}$ is a syntax splitting for $\rho^*(\preceq)$. □

In summary, we observe that the postulates (Cond_{coh}), (Marg_{coh}), $(\text{SynSplit}^b_{coh})$, and (SynSplit_{coh}) are fulfilled simultaneously by the epistemic state mapping ρ^* from Definition 14. In contrast to the stronger postulates (Cond), (Marg), (SynSplit^b), and (SynSplit) these postulates require certain sets of worlds to be coherent in the considered TPO in order to be applicable.

7 Conclusions and Future Work

In this article, we introduced the notion of epistemic state mappings, i.e., mappings among OCFs and TPOs. We proposed postulates for epistemic state mappings that ensure the preservation of certain properties of the epistemic state across the mapping. The properties considered in this article include the set of entailed conditionals and syntax splitting. Other postulates ensure compatibility with the operations marginalization and conditionalization. Furthermore, we investigated the relationships among the proposed postulates in general and for each combination of the considered framework. Some postulates are entailed by other postulates, e.g., (SynSplit) entails (SynSplit^b) and (IE) is equivalent to (Ord). We also showed that there are constellations and combinations of the postulates which cannot be satisfied simultaneously, e.g., there is no epistemic state mapping from TPOs to OCFs that fulfils both $(\text{wIE}^\Rightarrow)$ and (SynSplit^b). The only epistemic state mapping from TPOs to OCFs that fulfils both $(\text{wIE}^\Leftarrow)$ and (SynSplit^b) is the trivial mapping of every TPO to κ_{uni}, the state where every world is equally plausible. Using the notion of coherence, which is highly relevant e.g. for the concept of kinematics in belief revision [23, 32], we formulated weaker versions of the postulates for epistemic state mappings from TPOs to OCFs that avoid this incompatibility result.

Our current work includes extending the investigation of epistemic state mappings and their properties for establishing further relationships between OCFs and TPOs and thus to transfer more results between the two frameworks. Furthermore, we will consider epistemic state mappings among particular subclasses of TPOs and OCFs. We expect to find interesting and relevant subclasses such that epistemic state mappings over these subclasses fulfil combinations of postulates that are not fulfilled by epistemic state mappings over the full sets of TPOs and OCFs. Future work also includes using the insights of this paper to transfer tools and methods from the framework of TPOs to OCFs and vice versa.

Acknowledgments

This work was supported by the Deutsche Forschungsgemeinschaft (DFG, German Research Foundation), grant BE 1700/10-1 awarded to Christoph Beierle as part of the priority program "Intentional Forgetting in Organizations" (SPP 1921). Jonas Haldimann was supported by this grant.

References

[1] E.W. Adams. *The Logic of Conditionals*. D. Reidel, Dordrecht, 1975.

[2] Carlos E. Alchourrón, Peter Gärdenfors, and David Makinson. On the logic of theory change: Partial meet contraction and revision functions. *J. Symb. Log.*, 50(2):510–530, 1985.

[3] C. Beierle and G. Kern-Isberner. Semantical investigations into nonmonotonic and probabilistic logics. *Annals of Mathematics and Artificial Intelligence*, 65(2-3):123–158, 2012.

[4] Christoph Beierle, Tanja Bock, Gabriele Kern-Isberner, Marco Ragni, and Kai Sauerwald. Kinds and aspects of forgetting in common-sense knowledge and belief management. In Frank Trollmann and Anni-Yasmin Turhan, editors, *KI 2018*, volume 11117 of *LNAI*, pages 366–373. Springer, 2018.

[5] Christoph Beierle, Gabriele Kern-Isberner, Kai Sauerwald, Tanja Bock, and Marco Ragni. Towards a general framework for kinds of forgetting in common-sense belief management. *KI – Künstliche Intelligenz*, 33(1):57–68, 2019.

[6] S. Benferhat, D. Dubois, and H. Prade. Possibilistic and standard probabilistic semantics of conditional knowledge bases. *J. of Logic and Computation*, 9(6):873–895, 1999.

[7] Thomas Caridroit, Sébastien Konieczny, and Pierre Marquis. Contraction in propositional logic. *Int. J. Approx. Reason.*, 80:428–442, 2017.

[8] B. de Finetti. La prévision, ses lois logiques et ses sources subjectives. *Ann. Inst. H. Poincaré*, 7(1):1–68, 1937. Engl. transl. *Theory of Probability*, J. Wiley & Sons, 1974.

[9] James P. Delgrande. A knowledge level account of forgetting. *J. Artif. Intell. Res.*, 60:1165–1213, 2017.

[10] D. Dubois and H. Prade. Conditional objects as non-monotonic consequence relationships. *IEEE Transactions on Systems, Man, and Cybernetics*, 24(12):1724–1740, 1994.

[11] Thomas Eiter and Gabriele Kern-Isberner. A brief survey on forgetting from a knowledge representation and reasoning perspective. *KI – Künstliche Intelligenz*, 33(1):9–33, 2019.

[12] M. Goldszmidt and J. Pearl. Qualitative probabilities for default reasoning, belief revision, and causal modeling. *Artificial Intelligence*, 84:57–112, 1996.

[13] Jonas Haldimann, Christoph Beierle, and Gabriele Kern-Isberner. Postulates for transformations among epistemic states represented by ranking functions or total preorders. In Leila Amgoud and Richard Booth, editors, *19th International Workshop on Non-Monotonic Reasoning (NMR-2021)*, pages 213–222, 2021.

[14] Jonas Haldimann, Christoph Beierle, and Gabriele Kern-Isberner. Syntax splitting for iterated contractions, ignorations, and revisions on ranking functions using selection

strategies. In Wolfgang Faber, Gerhard Friedrich, Martin Gebser, and Michael Morak, editors, *Logics in Artificial Intelligence - 17th European Conference, JELIA 2021, Virtual Event, May 17-20, 2021, Proceedings*, volume 12678 of *Lecture Notes in Computer Science*, pages 85–100. Springer, 2021.

[15] Jonas Haldimann and Gabriele Kern-Isberner. On properties of epistemic state mappings among ranking functions and total preorders. In Christoph Beierle, Marco Ragni, Frieder Stolzenburg, and Matthias Thimm, editors, *Proceedings of the 7th Workshop on Formal and Cognitive Reasoning co-located with the 44th German Conference on Artificial Intelligence (KI 2021), September 28, 2021*, volume 2961 of *CEUR Workshop Proceedings*, pages 34–47. CEUR-WS.org, 2021.

[16] Jonas Philipp Haldimann, Gabriele Kern-Isberner, and Christoph Beierle. Syntax splitting for iterated contractions. In Diego Calvanese, Esra Erdem, and Michael Thielscher, editors, *Proceedings of the 17th International Conference on Principles of Knowledge Representation and Reasoning, KR 2020, Rhodes, Greece, September 12-18, 2020*, pages 465–475, 2020.

[17] Hirofumi Katsuno and Alberto O. Mendelzon. Propositional knowledge base revision and minimal change. *Artif. Intell.*, 52(3):263–294, 1992.

[18] Gabriele Kern-Isberner. A thorough axiomatization of a principle of conditional preservation in belief revision. *Annals of Mathematics and Artificial Intelligence*, 40(1-2):127–164, 2004.

[19] Gabriele Kern-Isberner. Axiomatizing a qualitative principle of conditional preservation for iterated belief change. In M. Thielscher, F. Toni, and F. Wolter, editors, *Principles of Knowledge Representation and Reasoning: Proceedings of the Sixteenth International Conference, KR 2018*, pages 248–256. AAAI Press, 2018.

[20] Gabriele Kern-Isberner, Christoph Beierle, and Gerhard Brewka. Syntax splitting = relevance + independence: New postulates for nonmonotonic reasoning from conditional belief bases. In Diego Calvanese, Esra Erdem, and Michael Thielscher, editors, *Principles of Knowledge Representation and Reasoning: Proceedings of the 17th International Conference, KR 2020*, pages 560–571. IJCAI Organization, 2020.

[21] Gabriele Kern-Isberner, Tanja Bock, Kai Sauerwald, and Christoph Beierle. Iterated contraction of propositions and conditionals under the principle of conditional preservation. In *GCAI-2017*, volume 50 of *EPiC Series in Computing*, pages 78–92. EasyChair, 2017.

[22] Gabriele Kern-Isberner and Gerhard Brewka. Strong syntax splitting for iterated belief revision. In *Proceedings of the Twenty-Sixth International Joint Conference on Artificial Intelligence, IJCAI 2017, Melbourne, Australia, August 19-25, 2017*, pages 1131–1137, 2017.

[23] Gabriele Kern-Isberner, Meliha Sezgin, and Christoph Beierle. A kinematics principle for iterated revision. *Artif. Intell.*, 314:103827, 2023.

[24] S. Kraus, D. Lehmann, and M. Magidor. Nonmonotonic reasoning, preferential models and cumulative logics. *Artificial Intelligence*, 44:167–207, 1990.

[25] Steven Kutsch. InfOCF-Lib: A Java library for OCF-based conditional inference. In

Christoph Beierle, Marco Ragni, Frieder Stolzenburg, and Matthias Thimm, editors, *Proceedings of the 8th Workshop on Dynamics of Knowledge and Belief (DKB-2019) and the 7th Workshop KI & Kognition (KIK-2019) co-located with 44nd German Conference on Artificial Intelligence (KI 2019), Kassel, Germany, September 23, 2019*, volume 2445 of *CEUR Workshop Proceedings*, pages 47–58. CEUR-WS.org, 2019.

[26] Steven Kutsch and Christoph Beierle. InfOCF-Web: An online tool for nonmonotonic reasoning with conditionals and ranking functions. In Zhi-Hua Zhou, editor, *Proceedings of the Thirtieth International Joint Conference on Artificial Intelligence, IJCAI 2021, Virtual Event / Montreal, Canada, 19-27 August 2021*, pages 4996–4999. ijcai.org, 2021.

[27] D. Lewis. *Counterfactuals*. Harvard University Press, Cambridge, Mass., 1973.

[28] Rohit Parikh. Beliefs, belief revision, and splitting languages. *Logic, Language, and Computation*, 2:266–278, 1999.

[29] J. Pearl. Bayesian and belief-functions formalisms for evidential reasoning: a conceptual analysis. In Z.W. Ras and M. Zemankova, editors, *Intelligent Systems – State of the art and future directions*, pages 73–117. Ellis Horwood, Chichester, 1990.

[30] Pavlos Peppas, Mary-Anne Williams, Samir Chopra, and Norman Y. Foo. Relevance in belief revision. *Artificial Intelligence*, 229((1-2)):126–138, 2015.

[31] Meliha Sezgin and Gabriele Kern-Isberner. Generalized ranking kinematics for iterated belief revision. In Roman Barták and Eric Bell, editors, *Proceedings of the Thirty-Third International FLAIRS Conference, FLAIRS-33*, pages 587–592. AAAI Press, 2020.

[32] Meliha Sezgin, Gabriele Kern-Isberner, and Christoph Beierle. Ranking kinematics for revising by contextual information. *Ann. Math. Artif. Intell.*, 89(10-11):1101–1131, 2021.

[33] W. Spohn. Ordinal conditional functions: a dynamic theory of epistemic states. In W.L. Harper and B. Skyrms, editors, *Causation in Decision, Belief Change, and Statistics, II*, pages 105–134. Kluwer Academic Publishers, 1988.

PROBABILISTIC DEONTIC LOGICS FOR REASONING ABOUT UNCERTAIN NORMS

VINCENT DE WIT
Department of Computer Science; University of Luxembourg, Luxembourg
vincent.j.wit@gmail.com

DRAGAN DODER
Department of Information and Computing Sciences; Utrecht University, The Netherlands
d.doder@uu.nl

JOHN JULES MEYER
Department of Information and Computing Sciences; Utrecht University, The Netherlands
J.J.C.Meyer@uu.nl

Abstract

In this article, we present a proof-theoretical and model-theoretical approach to probabilistic logic for reasoning about uncertainty about normative statements. We introduce two logics with languages that extend both the language of monadic deontic logic and the language of probabilistic logic. The first logic allows statements like "the probability that one is obliged to be quiet is at least 0.9". The second logic allows iteration of probabilities in the language. We axiomatize both logics, provide the corresponding semantics and prove that the axiomatizations are sound and complete. We also prove that both logics are decidable. In addition, we show that the problem of deciding satisfiability for the simpler of our two logics is in PSPACE, no worse than that of deontic logic.

Keywords: MDL; Normative reasoning; Probabilistic logic; Completeness; Decidability.

This paper is a revised and extended version of our conference paper [20] presented at the Sixteenth European Conference on Symbolic and Quantitative Approaches to Reasoning with Uncertainty (ECSQARU 2021), in which we introduced a logic for reasoning about probabilities of a deontic statement, provided a complete axiomatization for the logic and proved its decidability. In this paper, we extend those results, by providing the complexity result for the satisfiability problem. Additionally, we also present in this paper another, richer logic with nesting of probability operators. For that novel logic we also provide an axiomatization, prove its completeness, and show that the logic is decidable.

1 Introduction

Norms govern many parts of human life and these norms need to be learned at some point. This means that before a norm is learned there is uncertainty about whether a norm holds or not. To formalize this notion a probabilistic deontic logic is developed in this paper. The seminal work of von Wright from 1951 [21] initiated a systematic study on the formalization of normative reasoning in terms of deontic logic. The latter is a branch of modal logic that deals with obligation, permission, and related normative concepts. A plethora of deontic logics have been developed for various application domains like legal reasoning, argumentation theory, and normative multi-agent systems [1, 5, 11].

Some recent work studied learning of behavioral norms from data [16, 18]. In [16] the authors pointed out that human norms are context-specific and laced with uncertainty, which poses challenges to their representation, learning, and communication. They gave an example of a learner that might conclude from observations that talking is prohibited in a library setting, while another learner might conclude the opposite when seeing people talking at the checkout counter. They represented uncertainty about norms using deontic operators equipped with probabilistic boundaries that capture the subjective degree of certainty.

In this paper, we study uncertain norms from a logical point of view. We use probabilistic logic [6, 7, 8, 10, 17, 19] to represent uncertainty, and we present the proof-theoretical and model-theoretical approach to a logic which allows reasoning about uncertain normative statements. We take two well-studied logics, monadic deontic logic (MDL) [14] and probabilistic logic of Fagin, Halpern, and Magido (FHM) [7], as the starting point, and combine them in a rich formalism that generalizes each of them. The resulting language makes it possible to adequately model different degrees of belief in norms; for example, we can express statements like "the probability that one is obliged to be quiet is at least 0.9".

The semantics for our main logic $\mathcal{PMDL}$ consists of specific Kripke-like structures, where each model contains a probability space whose sample space is the set of states, and with each state carrying enough information to evaluate a deontic formula. We consider so-called *measurable* models, which allow us to assign a probability value to every deontic statement. We also propose another, richer language $\mathcal{PMDL}^2$ which allows nesting of probability operators. In this case, the semantics is naturally generalized i.e. that all states are equipped with probability spaces. In addition to the nesting of operators, we modify the language in such a way that we allow different agents to place (possibly different) probabilities on norms and events. So the formulas can express one's uncertainty about another person's uncertainty about norms.

The main result of this article is a sound and complete axiomatization for our logics. Like any other real-valued probabilistic logic, it is not compact, so any finitary axiomatic system would fail to be strongly complete ("every consistent set of formulas has a model") [10]. We prove weak completeness ("every consistent formula has a model") by combining and modifying completeness techniques for MDL and FHM. We also show that our logics are decidable; combining the method of filtration and a reduction to a system of inequalities. In addition, we show that the problem of deciding satisfiability for the logic $\mathcal{PMDL}$ is in PSPACE, no worse than that of deontic logic.

The rest of the paper is organized as follows: In Section 2, the proposed syntax and semantics of the logic will be presented together with other needed definitions. In Section 3, the axiomatization of the logic is given; in section 4, soundness and completeness is proven. In Section 5, we show that our logic is decidable; in Section 6, the probability structure of the logic is changed such that iterations of probabilities are possible, and completeness and decidability is proven. Lastly, in Section 7, a conclusion is given together with future research topics.

2 Syntax and Semantics

In this section, we present the syntax and semantics of our probabilistic deontic logic. This logic, which we named $\mathcal{PMDL}$, contains two types of formulas: standard deontic formulas of MDL, and probabilistic formulas. Let $\mathbb{N}$ denote the set of integers.

Definition 1 (Formula). *Let $\mathbb{P}$ be a set of atomic propositions. The language $\mathcal{L}$ of probabilistic MDL is generated by the following two sentences of BNF (Backus Naur Form):*

$$[\mathcal{L}_{MDL}] \quad \phi ::= p \mid \neg \phi \mid (\phi \wedge \phi) \mid O\phi \quad p \in \mathbb{P}$$
$$[\mathcal{L}_{\mathcal{PMDL}}] \quad f ::= a_1 w(\phi_1) + \cdots + a_n w(\phi_n) \geq \alpha \mid \neg f \mid (f \wedge f) \quad a_i, \alpha \in \mathbb{N}$$

The set of all formulas $\mathcal{L}$ is $\mathcal{L}_{MDL} \cup \mathcal{L}_{\mathcal{PMDL}}$. We denote the elements of $\mathcal{L}$ with θ and θ', possibly with subscripts.

The construct $O\phi$ reads as "It is obligatory that ϕ", while $w(\phi)$ stands for "probability of ϕ". An expression of the form $a_1 w(\phi_1) + \cdots + a_n w(\phi_n)$ is called *term*. We denote terms with x and t, possibly with subscripts. The propositional connectives, $\vee, \rightarrow$ and $\leftrightarrow$, are introduced as abbreviations, in the usual way. There are also two additional deontic operators that denote the following: forbidden, $F\phi \equiv O\neg\phi$; and permitted $P\phi \equiv \neg F\phi \wedge \neg O\phi$. We abbreviate $\theta \wedge \neg \theta$ with $\bot$, and $\neg\bot$ with $\top$. We also use abbreviations to define other types of inequalities; for example, $w(\phi) \geq w(\phi')$ is an abbreviation for $w(\phi) - w(\phi') \geq 0$, $w(\phi) = \alpha$ for $w(\phi) \geq \alpha$ and $-w(\phi) \geq -\alpha$, $w(\phi) < \alpha$ for $\neg w(\phi) \geq \alpha$.

It is very important to mention that we can also use abbreviations that allow us to see *rational numbers* as coefficients of terms i.e. they can be eliminated from any formula by clearing the denominator. For example, the formula

$$\frac{2}{3}t_1 + \frac{3}{4}t_2 \geq 1$$

is simply an abbreviation for $8t_1 + 9t_2 \geq 12$.

Example 1. *Following our informal example from the introduction about behavioral norms in a library, the fact that a person has become fairly certain that it is normal to be quiet might be expressed by the probabilistic statement "the probability that one is obliged to be quiet is at least 0.9". This sentence could be formalized using the introduced language as*

$$w(Oq) \geq 0.9.$$

Note that we do not allow mixing of the formulas from $\mathcal{L}_{MDL}$ and $\mathcal{L}_{PMDL}$. For example, $O(p \vee q) \wedge w(Oq) \geq 0.9$ is not a formula of our language. Before we introduce the semantics of $\mathcal{PMDL}$ we will introduce MDL models.

Definition 2 (MDL model). *An MDL model D is a tuple $D = (W, R, V)$ where:*

- *W is a (non-empty) set of "possible worlds"; W is called the universe of the model.*

- *$R \subseteq W \times W$ is a binary relation over W, such that*

$$(\forall w \in W)(\exists u \in W)(wRu). \qquad \text{(seriality)}$$

If $(w, u) \in R$, we say that u is an R–successor of w.

- *$V : \mathbb{P} \to 2^W$ is a valuation function assigning to each atom p a set $V(p) \subseteq W$ (intuitively the set of worlds at which p is true.)*

We denote the set of all MDL models with $\mathbb{D}$. As formalized in the following definition, the relation R relates worlds to worlds, with the intention that everything obligatory at a world holds in its R successors.

Next, we define the satisfiability relation of MDL.

Definition 3 (Satisfaction in MDL). *Let $D = (W, R, V)$ be an MDL model, and let $w \in W$. We define the satisfiability of a deontic formula $\phi \in \mathcal{L}_{MDL}$ in the world w, denoted by $D, w \models_{MDL} \phi$, recursively as follows:*

- $D, w \models_{MDL} p$ iff $w \in V(p)$.
- $D, w \models_{MDL} \neg \phi$ iff $D, w \not\models_{MDL} \phi$.
- $D, w \models_{MDL} \phi \wedge \psi$ iff $D, w \models_{MDL} \phi$ and $D, w \models_{MDL} \psi$.
- $D, w \models_{MDL} O\phi$ iff for all $u \in W$, if wRu then $D, u \models_{MDL} \phi$.

Now we introduce the semantics of $\mathcal{PMDL}$.

Definition 4 ($\mathcal{PMDL}$ Model). *A probabilistic deontic model is a tuple $M = \langle S, \mathscr{X}, \mu, \tau \rangle$, where*

- *S is a non-empty set of states*
- *$\mathscr{X}$ is a σ-algebra of subsets of S*
- *$\mu : \mathscr{X} \to [0, 1]$ is a probability measure, i.e.,*
    - *$\mu(X) \geq 0$ for all $X \in \mathscr{X}$*
    - *$\mu(S) = 1$*
    - *$\mu(\bigcup_{i=1}^{\infty} X_i) = \sum_{i=1}^{\infty} \mu(X_i)$, if the X_i's are pairwise disjoint members of $\mathscr{X}$*
- *τ is a function that assigns to each state in S a pair consisting of an MDL model and a world of that model, i.e., $\tau : s \mapsto (D_s, w_s)$, where:*
    - *$D_s = (W_s, R_s, V_s) \in \mathbb{D}$*
    - *$w_s \in W_s$*

Let us illustrate this definition.

Example 1. *(continued)* Assume a finite set of atomic propositions $\{p, q\}$. Let us consider the model $M = \langle S, \mathscr{X}, \mu, \tau \rangle$, where

- $S = \{s_1, s_2, s_3, s_4\}$
- $\mathscr{X}$ is the set of all subsets of S

- μ is characterized by: $\mu(\{s_1\}) = 0.5$, $\mu(\{s_2\}) = \mu(\{s_3\}) = 0.2$, $\mu(\{s_4\}) = 0.1$ (other values follow from the properties of probability measures)

- τ is a mapping which assigns to the state s_1, $D_{s_1} = (W_{s_1}, R_{s_1}, V_{s_1})$ and w_{s_1} such that

    - $W_{s_1} = \{w_1, w_2, w_3, w_4\}$
    - $R_{s_1} = \{(w_1, w_2), (w_1, w_3), (w_2, w_2), (w_2, w_3), (w_3, w_2), (w_3, w_3), (w_4, w_2),$ $(w_4, w_3), (w_4, w_4)\}$
    - $V_{s_1}(p) = \{w_1, w_3\}$, $V_s(q) = \{w_2, w_3\}$
    - $w_{s_1} = w_1$

    Note that the domain of τ is always the whole set S, but in this example we only explicitly specify $\tau(s_1)$ for illustration purposes.

This model is depicted in Figure 1. The circle on the right contains the four states of the model, which are measured by μ. Each of the states is equipped with a standard pointed model of MDL. In this picture, only one of them is shown, the one that corresponds to s_1. It is represented within the circle on the left. Note that the

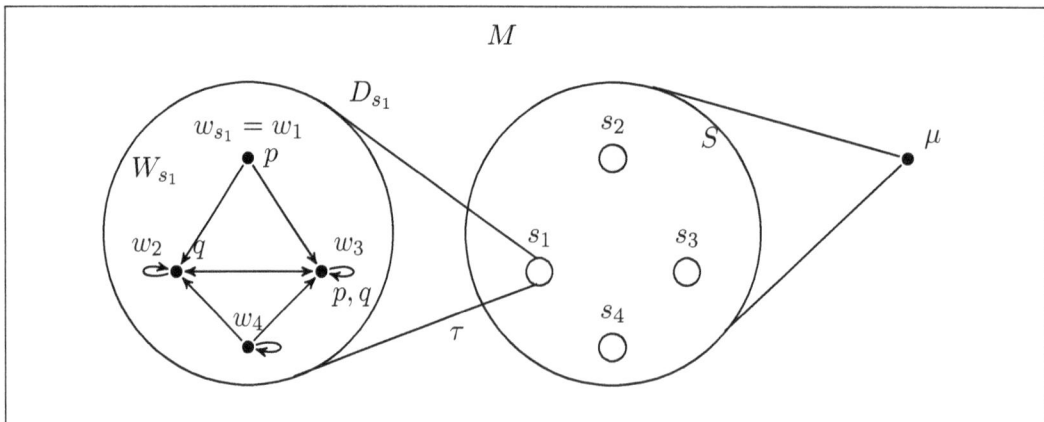

Figure 1: Model $M = \langle S, \mathcal{X}, \mu, \tau \rangle$.

arrows depict the relation R. If we assume that q stands for "quiet", like in the previous example, in all R-successors of w_1 the proposition q holds. Note that, according to Definition 3, this means that in w_1 people are obliged to be quiet in the library.

For a model $M = \langle S, \mathcal{X}, \mu, \tau \rangle$ and a formula $\phi \in \mathcal{L}_{MDL}$, let $\|\phi\|_M$ denote the set of states that satisfy ϕ, i.e., $\|\phi\|_M = \{s \in S \mid D_s, w_s \models_{MDL} \phi\}$. We omit the subscript M from $\|\phi\|_M$ when it is clear from context. The following definition introduces an important class of probabilistic deontic models, the so-called *measurable* models.

Definition 5 (Measurable model). *A probabilistic deontic model is measurable if*

$$\|\phi\|_M \in \mathscr{X}$$

for every $\phi \in \mathcal{L}_{MDL}$. Denote the class of all measurable models of $\mathcal{PMDL}$ by $\mathcal{PMDL}^{Meas}$.

In this paper, we focus on measurable structures, and we prove soundness & completeness, and decidability results for this class of structures.

Definition 6 (Satisfaction). *Let $M = \langle S, \mathscr{X}, \mu, \tau \rangle \in \mathcal{PMDL}^{Meas}$ be a measurable probabilistic deontic model. We define the satisfiability relation $\models$ recursively as follows:*

- $M \models \phi$ iff $D_s, w_s \models_{MDL} \phi$ holds for every $s \in S$, where $\tau(s) = (D_s, w_s)$
- $M \models a_1 w(\phi_1) + \cdots + a_k w(\phi_k) \geq \alpha$ iff $a_1 \mu(\|\phi_1\|) + \cdots + a_k \mu(\|\phi_k\|) \geq \alpha$.
- $M \models \neg f$ iff $M \not\models f$
- $M \models f \wedge g$ iff $M \models f$ and $M \models g$

Example 1. *(continued) Continuing the previous example, it is now also possible to speak about the probability of the obligation to be quiet in a library. First, according to Definition 3 it holds that $D_{s_1}, w_{s_1} \models_{MDL} Oq$. Furthermore, assume that τ is defined in the way such that $D_{s_2}, w_{s_2} \models_{MDL} Oq$ and $D_{s_3}, w_{s_4} \models_{MDL} Oq$, but $D_{s_4}, w_{s_4} \not\models_{MDL} Oq$. Then $\mu(\|Oq\|) = \mu(\{s_1, s_2, s_3\}) = 0.5 + 0.2 + 0.2 = 0.9$. According to Definition 6, $M \models w(Oq) \geq 0.9$.*

Note that, according to Definition 6, a deontic formula is true in a model iff it holds in every state of the model. This is a consequence of our design choice that those formulas represent undisputable deontic knowledge, while probabilistic formulas express uncertainty about norms. At the end of this section, we define some standard semantical notions.

Definition 7 (Semantical consequence). *Given a set Γ of formulas, a formula θ is a semantical consequence of Γ (notation: $\Gamma \models \theta$) whenever all the states of the model have, if $M, s \models \theta'$ for all $\theta' \in \Gamma$, then $M, s \models \theta$.*

Definition 8 (Validity). *A formula θ is valid (notations: $\models \theta$) whenever for $M = \langle S, \mathscr{X}, \mu, \tau \rangle$ and every $s \in S$: $M, s \models \theta$ holds.*

3 Axiomatization

The following axiomatization contains 13 axioms and 3 inference rules. It combines the axioms of proof system D of MDL [14] with the axioms of probabilistic logic. The axioms for reasoning about linear inequalities are taken from [7].

The Axiomatic System: $AX_{\mathcal{PMDL}}$

Tautologies and Modus Ponens

Taut. All instances of propositional tautologies.

MP. From θ and $\theta \to \theta'$ infer θ'.

Reasoning with O:

O-K. $O(\phi \to \psi) \to (O\phi \to O\psi)$

O-D. $O\phi \to P\phi$

O-Nec. *From ϕ infer $O\phi$*

Reasoning about linear inequalities:

I1. $x \geq x$ (identity)

I2. $(a_1 x_1 + ... + a_k x_k \geq c) \leftrightarrow (a_1 x_1 + ... + a_k x_k + 0 x_{k+1} \geq c)$ (adding and deleting 0 terms)

I3. $(a_1 x_1 + ... + a_k x_k \geq c) \to (a_{j_1} x_{j_1} + ... + a_{j_k} x_{j_k} \geq c)$, if $j_1, ..., j_k$ is a permutation of $1, ..., k$ (permutation)

I4. $(a_1 x_1 + ... + a_k x_k \geq c) \wedge (a'_1 x_1 + ... + a'_k x_k \geq c') \to ((a_1 + a'_1) x_1 + ... + (a_k + a'_k) x_k \geq (c + c'))$ (addition of coefficients)

I5. $(a_1 x_1 + ... + a_k x_k \geq c) \leftrightarrow (d a_1 x_1 + ... + d a_k x_k \geq dc)$ if $d > 0$ (multiplication of non-zero coefficients)

I6. $(t \geq c) \vee (t \leq c)$ if t is a term (dichotomy)

I7. $(t \geq c) \to (t > d)$ if t is a term and $c > d$ (monotonicity)

Reasoning about probabilities:

W1. $w(\phi) \geq 0$ (non-negativity).

W2. $w(\phi \vee \psi) = w(\phi) + w(\psi)$, if $\neg(\phi \wedge \psi)$ is an instance of a classical propositional tautology (finite additivity).

W3. $w(\top) = 1$

P-Dis. From $\phi \leftrightarrow \psi$ infer $w(\phi) = w(\psi)$ (probabilistic distributivity)

The axiom Taut allows all $\mathcal{L}_{MDL}$-instances and $\mathcal{L}_{PMDL}$-instances of propositional tautologies. For example, $w(Oq) \geq 0.9 \vee \neg w(Oq) \geq 0.9$ is an instance of Taut, but $w(Oq) \geq 0.9 \vee \neg w(Oq) \geq 1$ is not. Note that Modus Ponens (MP) can be applied to both types of formulas, but only if θ and θ' are both from $\mathcal{L}_{MDL}$ or both from $\mathcal{L}_{PMDL}$. O-Nec is a deontic variant of necessitation rule. P-Dis is an inference rule which states that two equivalent deontic formulas must have the same probability values.

Definition 9 (Syntactical consequence). *A derivation of θ is a finite sequence $\theta_1, \ldots, \theta_m$ of formulas such that $\theta_m = \theta$, and every θ_i is either an instance of an axiom, or it is obtained by the application of an inference rule to formulas in the sequence that appear before θ_i. If there is a derivation of θ, we say that θ is a theorem and write $\vdash \theta$. We also say that θ is derivable from a set of formulas Γ, and write $\Gamma \vdash \theta$, if there is a finite sequence $\theta_1, \ldots, \theta_m$ of formulas such that $\theta_m = \theta$, and every θ_i is either a theorem, a member of Γ or the result of an application of MP or P-Dis to formulas in the sequence that appear before θ_i.*

Note that this definition restricts the application of O-Nec. to theorems only. This is a standard restriction for modal necessitations, which enables one to prove the Deduction Theorem using induction on the length of the inference. Also, note that only deontic formulas can participate in a proof of another deontic formula, thus derivations of deontic formulas in our logic coincide with their derivations in MDL.

Definition 10 (Consistency). *A set Γ is consistent if $\Gamma \nvdash \bot$, and inconsistent otherwise.*

Now we prove some basic consequences of AX_{PMDL}. The first one is the probabilistic variant of necessitation. It captures the semantical property that a deontic formula represents undisputable knowledge, and therefore it must have a probability value of 1. The second point states that we can derive from the axiomatization that the weight of falsum equals zero. The third part of the lemma shows that a form of additivity proposed as an axiom in [7] is provable in AX_{PMDL}.

Lemma 1. *The following rules are derivable from our axiomatization:*

1. *From ϕ infer $w(\phi) = 1$*
2. *$\vdash w(\bot) = 0$*
3. *$\vdash w(\phi \wedge \psi) + w(\phi \wedge \neg\psi) = w(\phi)$.*

Proof.

1. Let us assume that a formula ϕ is derived. Then, using propositional reasoning (Taut and MP), one can infer $\phi \leftrightarrow \top$. Consequently, $w(\phi) = w(\top)$ follows from the rule P-Dis. Since we have that $w(\top) = 1$ (by W3), we can employ the axioms for reasoning about inequalities to infer $w(\phi) = 1$.

2. Then to show that $w(\bot) = 0$ using finite additivity (W2) $w(\top \vee \neg\top) = w(\top) + w(\neg\top) = 1$ and so $w(\neg\top) = 1 - w(\top)$. Since $w(\top) = 1$ and $\neg\top \leftrightarrow \bot$ we can derive $w(\bot) = 0$.

3. To derive additivity we begin with the propositional tautology, $\neg((\phi \wedge \psi) \wedge (\phi \wedge \neg\psi))$ then the following equation is given by W2 $w(\phi \wedge \psi) + w(\phi \wedge \neg\psi) = w((\phi \wedge \psi) \vee (\phi \wedge \neg\psi))$. The disjunction $(\phi \wedge \psi) \vee (\phi \wedge \neg\psi)$ can be rewritten to, $\phi \wedge (\psi \vee \neg\psi)$ which is equivalent to ϕ. From $\phi \leftrightarrow (\phi \wedge \psi) \vee (\phi \wedge \neg\psi)$, using P-Dis, we obtain $w(\phi) = w(\phi \wedge \psi) + w(\phi \wedge \neg\psi)$.

□

4 Soundness & Completeness

In this section, we prove that our logic is sound and complete with respect to the class of measurable models; combining, adapting and following the approaches from [7, 3].

Theorem 1 (Soundness & Completeness). *The axiom system $AX_{\mathcal{PMDL}}$ is sound and complete with respect to the class of measurable models $\mathcal{PMDL}^{Meas}$, i.e., $\vdash \theta$ iff $\models \theta$.*

Proof. The proof of soundness is straightforward. To prove completeness, we need to show that every consistent formula θ is satisfied in a measurable model. Since we have two types of formulas, we distinguish two cases.

If $\theta \in \mathcal{L}_{MDL}$ we write θ as ϕ. Since ϕ is consistent and MDL is complete [14], we know that there is an MDL model (W, R, V) and $w \in W$ such that $(W, R, V), w \models \phi$.

Then, for any probabilistic deontic model M with only one state s and $\tau(s) = ((W, R, V), w)$ we have $M, s \models \phi$, and therefore $M \models \phi$ (since s is the only state); so the formula is satisfiable.

When $\theta \in \mathcal{L}_{\mathcal{PMDL}}$ we write θ as f. Then f is consistent and we prove that f is satisfiable. First notice that f can be equivalently rewritten as a formula in disjunctive normal form,

$$f \leftrightarrow g_1 \vee \cdots \vee g_n \tag{1}$$

this means that satisfiability of f can be proven by showing that one of the disjuncts g_i of the disjunctive normal form of f is satisfiable. Note that every disjunct is of the form:

$$g_i = \bigwedge_{j=1}^{r} (\sum_k a_{j,k} w(\phi_{j,k}) \geq c_j) \wedge \bigwedge_{j=r+1}^{r+s} \neg (\sum_k a_{j,k} w(\phi_{j,k}) \geq c_j). \tag{2}$$

To show that g_i is satisfiable we will substitute each weight term $w(\phi_{j,k})$ by a sum of weight terms that take as arguments formulas from the set Δ that will be constructed below. For any formula θ, let us denote the set of subformulas of θ by $Sub(\theta)$. Then, for considered, g_i we introduce the set of all deontic subformulas $Sub_{DL}(g_i) = Sub(g_i) \cap \mathcal{L}_{MDL}$. We create the set Δ as the set of all possible formulas that are conjunctions of formulas from $Sub_{DL}(g_i) \cup \{\neg e \mid e \in Sub_{DL}(g_i)\}$, such that for every e either e or $\neg e$ is taken as a conjunct (but not both). Then we can prove the following two claims about the set Δ:

- The conjunction of any two different formulas δ_k and δ_l from Δ is inconsistent: $\vdash \neg(\delta_k \wedge \delta_l)$. This is the case because for each pair of δ's at least one subformula $e \in Sub_{DL}(g_i)$ such that $\delta_k \wedge \delta_l \vdash e \wedge \neg e$ and $e \wedge \neg e \vdash \bot$. If there is no such e, then by construction $\delta_k = \delta_l$.

- The disjunction of all δ's in Δ is a tautology: $\vdash \bigvee_{\delta \in \Delta} \delta$. Indeed, it is clear from the way the set Δ is constructed, that the disjunction of all formulas is an instance of a propositional tautology.

As noted earlier, we will substitute each term of each weight formula of g_i with a sum of weight terms. This can be done by using the just introduced set Δ and the set Φ, which we define as the set containing all deontic formulas $\phi_{j,k}$ that occur in the weight terms of g_i. In order to get all the relevant δ's to represent a weight term, we construct for each $\phi \in \Phi$ the set $\Delta_\phi = \{\delta \in \Delta \mid \delta \vdash \phi\}$ which contains all δ's that imply ϕ. Then we can derive the following equivalence:

$$\vdash \phi \leftrightarrow \bigvee_{\delta \in \Delta_\phi} \delta.$$

From the rule P-Dis we obtain

$$\vdash w(\phi) = w(\bigvee_{\delta \in \Delta_\phi} \delta).$$

Since any two elements of Δ are inconsistent, from W2 and axioms about inequalities we obtain

$$\vdash w(\bigvee_{\delta \in \Delta_\phi} \delta) = \sum_{\delta \in \Delta_\phi} w(\delta).$$

Consequently, we have

$$\vdash w(\phi) = \sum_{\delta \in \Delta_\phi} w(\delta). \tag{3}$$

Note that some of the formulas δ's might be inconsistent (for example, a formula from Δ might be a conjunction in which both $O(p \wedge q)$ and Fp appear as conjuncts). For an inconsistent formula δ, we have $\vdash \delta \leftrightarrow \bot$ and, consequently $\vdash w(\delta) = 0$, by the inference rule P-Dis. This provably filters out the inconsistent δ's from each weight formula, using the axioms about linear inequalities. Thus, without any loss of generality, we can assume in the rest of the proof that all the formulas from Δ are consistent[2].

Lets us consider a new formula f', created by substituting each term of each weight formula of g_i from (1), thus transforming each conjunct (2) into

$$g_i' = \left(\bigwedge_{j=1}^{r} (\sum_k a_{j,k} \sum_{\delta \in \Delta_{\phi_{j,k}}} w(\delta) \geq c_j) \right) \wedge \left(\bigwedge_{j=r+1}^{r+s} \neg (\sum_k a_{j,k} \sum_{\delta \in \Delta_{\phi_{j,k}}} w(\delta) \geq c_j) \right)$$

Since consistency of the formula f is equivalent to consistency of one of its disjuncts g_i from (1), in the rest of the proof we will focus on one such disjunct, g_i. Note that (3) implies that g_i and g_i' are two provably equivalent formulas (and the same holds for f and f'). Then we will construct g_i'' by adding to g_i': a non-negativity constraint and an equality that binds the total probability weight of δ's to 1. In other words, g_i'' is the conjunction of the following formulas:

---

[2] We might introduce Δ^c and Δ_ϕ^c as the sets of all consistent formulas from Δ and Δ_ϕ, respectively, but since we will still have $\vdash w(\phi) = \sum_{\delta \in \Delta_\phi^c} w(\delta)$, we prefer not to burden the notation with the superscripts in the rest of the proof, and we assume that we do not have inconsistent formulas in Δ.

$$\sum_{\delta \in \Delta} w(\delta) = 1$$

$\forall \delta \in \Delta \qquad\qquad\qquad\qquad\qquad\qquad\qquad w(\delta) \geq 0$

$$\forall l \in \{1, \ldots, r\} \qquad\qquad \sum_k a_{l,k} \sum_{\delta \in \Delta_{\phi_{l,k}}} w(\delta) \geq c_l$$

$$\forall l \in \{r+1, \ldots, r+s\} \qquad \sum_k a_{l,k} \sum_{\delta \in \Delta_{\phi_{l,k}}} w(\delta) < c_l$$

Since the weights can be attributed independently while respecting the system of equations and inequalities, the formula g_i'' is satisfiable iff the corresponding system of equations and inequalities, that we denote by $Sys(g_i'')$ is solvable:

$$\sum_{i=1}^{|\Delta|} x_i = 1$$

$\forall i \in \{1, \ldots, |\Delta|\} \qquad\qquad\qquad\qquad\qquad x_i \geq 0$

$$\forall l \in \{1, \ldots, r\} \qquad\qquad \sum_k a_{l,k} \sum_{i=1}^{|\Delta_{\phi_{l,k}}|} x_i \geq c_l$$

$$\forall l \in \{r+1, \ldots, r+s\} \qquad \sum_k a_{l,k} \sum_{i=1}^{|\Delta_{\phi_{l,k}}|} x_{r+i} < c_l$$

Initially we considered a consistent formula g_i and transformed it to a provably equivalent formula g_i''. Proving satisfiability of g_i'' is equivalent to proving satisfiability of g_i; since the set of models of g_i'' coincides with the set of models of g_i', which in turn has the same models as g_i.

Using proof from the incongruous we assume g_i'' to be unsatisfiable and show that this leads to a contradiction. Since g_i'' is assumed unsatisfiable this means that the system of linear inequalities $Sys(g_i'')$ does not have a solution. This further means that in the process of solving the system $Sys(g_i'')$ (using any procedure for solving linear inequalities, e.g. we can use Fourier–Motzkin elimination) we would obtain an equivalent system containing an equation or inequality without solutions. Without any loss of generality, assume that the obtained formula is $0 = 1$. Now, since we have the axioms I1-I7 as a part of our $AX_{\mathcal{PMDL}}$, we can "syntactically" derive all those corresponding steps (of transforming inequalities using the procedure for solving linear inequalities) from g_i'' using our axiomatization, and therefore we obtain that $0 = 1$ is a formula (of our logic $\mathcal{PMDL}$) that can be derived from g_i''. That means g_i'' is inconsistent, which is a contradiction because we started with g_i as a consistent formula. $\square$

5 Decidability

In this section, we prove that our logic $\mathcal{PMDL}$ is decidable, and we show that there is a decidability procedure for the problem that runs in polynomial space. First, let us recall the satisfiability problem: given a formula θ, we want to determine if there exists a model M such that $M \models \theta$.

Theorem 2 (Decidability). *Satisfiability problem for $\mathcal{PMDL}$ is decidable.*

Proof. Since we have two types of formulas, we will consider two cases. First, let us assume that $\theta \in \mathcal{L}_{MDL}$. We start with the well-known result that the problem of whether a formula from $\mathcal{L}_{MDL}$ is satisfiable in an MDL model is decidable [14]. It is sufficient to show that each $\theta \in \mathcal{L}_{MDL}$ is satisfiable in an MDL model iff it is satisfiable under our semantics. First, if $(W', R', V'), w' \models \theta$ for some deontic model (W', R', V') and $w' \in W'$, let us construct the model $M = \langle S, \mathscr{X}, \mu, \tau \rangle$, with $S = \{s\}$, $\mathscr{X} = \{\emptyset, S\}$, $\mu(S) = 1$ and $\tau(s) = ((W', R', V'), w')$. Since $(W', R', V'), w' \models \theta$, then $M, s \models \theta$. From the fact that s is the unique state of M, we conclude that $M \models \theta$. On the other hand, if θ is not satisfiable in MDL, then for every $M = \langle S, \mathscr{X}, \mu, \tau \rangle$ and $s \in S$ we will have $M, s \not\models \theta$, so $M \not\models \theta$.

Now, let us consider the case $\theta \in \mathcal{L}_{\mathcal{PMDL}}$. In the proof, we use the method of filtration [12, 3], and reduction to finite systems of inequalities. We only provide a sketch of the proof, since we use similar ideas as in our completeness proof. We will also use the notation introduced in the proof of completeness. In the first part of the proof, we show that a formula is satisfiable iff it is satisfiable in a model with a finite number of (1) states and (2) worlds.

(1) First we show that if $\theta \in \mathcal{L}_{\mathcal{PMDL}}$ is satisfiable, then it is satisfiable in a model with a finite set of states, whose size is at most $2^{|Sub_{DL}(\theta)|}$ (where $Sub_{DL}(\theta)$ is the set of deontic subformulas of θ, as defined in the proof of Theorem 1). Let $M = \langle S, \mathscr{X}, \mu, \tau \rangle$ be a model such that $M \models \theta$. Let us define by $\sim$ the equivalence relation over $S \times S$ in the following way: $s \sim s_2$ iff for every $\phi \in Sub_{DL}(\theta)$, $M, s \models \phi$ iff $s_2 \models \phi$. Then the corresponding quotient set $S_{/\sim}$ is finite and $|S_{/\sim}| \leq 2^{|Sub_{DL}(\theta)|}$. Note that every C_i belongs to $\mathscr{X}$, since it corresponds to a formula δ_i of Δ (from the proof of Theorem 1), i.e., $C_i = \|\delta_i\|$. Next, for every equivalence class, C_i we choose one element and denote it s_i. Then we consider the model $M' = \langle s_2, \mathscr{X}', \mu', \tau' \rangle$, where:

- $s_2 = \{s_i \mid C_i \in S_{/\sim}\}$,

- $\mathscr{X}'$ is the power set of s_2,

- $\mu'(\{s_i\}) = \mu(C_i)$ such that $s_i \in C_i$ and for any $X \subseteq s_2$, $\mu'(X) = \sum_{s_i \in X} \mu'(\{s_i\})$,

- $\tau'(s_i) = \tau(s_i)$.

Then it is straightforward to verify that $M' \models \theta$. Moreover, note that, by definition of M', for every $s_i \in s_2$ there is $\delta_i \in \Delta$ such that $M', s_i \models \delta_i$, and that for every $s_j \neq s_i$ we have $M', s_j \not\models \delta_i$. We therefore say that δ_i is the *characteristic formula* of s_i.

(2) Even if s_2 is finite, some sets of worlds attached to a state might be infinite. Now we will modify τ', to ensure that every $W(s_i)$ is finite, and of the size which is bounded by a number that depends on the size of θ. In this part of the proof, we refer to the filtration method used to prove completeness of MDL [3], which shows that if a deontic formula ϕ is satisfiable, it is satisfied in a world of a model $D(\psi) = (W, R, V)$ where the size of W is at most exponential wrt. the size of the set of subformulas of ϕ. Then we can replace τ' with a function τ'' which assigns to each s_i one such $D(\delta_i)$ and the corresponding world, where δ_i is the characteristic formula of s_i. We also assume that each $V(s_i)$ is restricted to the propositional letters from $Sub_{DL}(\theta)$. Finally, let $M'' = \langle s_2, \mathscr{X}', \mu', \tau'' \rangle$ It is easy to check that for every $\phi \in Sub_{DL}(\theta)$ and $s_i \in s_2$, $M', s_i \models \phi$ iff $M'', s_i \models \phi$. Therefore, $M'' \models \theta$.

From the steps (1) and (2) it follows that in order to check if a formula $\theta \in \mathcal{L}_{\mathcal{PMDL}}$ is satisfiable, it is enough to check if it is satisfied in a model $M = \langle S, \mathscr{X}, \mu, \tau \rangle$ in which S and each W_s (for every $s \in S$) are of finite size, bounded from above by a fixed number depending on the size of $|Sub_{DL}(\theta)|$. Then there are finitely many options for the choice of S and τ (i.e., (D_s, w_s), for every $s \in S$), and our procedure can determine in finite time whether there is a probability measure μ for some of them, such that θ holds in the model. We convert our formula f into a formula in the complete disjunctive form as in (1). We guess S and τ and check whether we can assign probability values to the states from S, considering each disjunct g_i and using translation to a system of linear inequalities, in the same way as we have done in the proof of Theorem 1. This finishes the proof since the problem of checking whether a linear system of inequalities has a solution is decidable. $\square$

Moreover, we show that there is a procedure that decides the satisfiability of any formula of $\mathcal{PMDL}$ in $\mathcal{PSPACE}$.

Theorem 3. *There is a procedure that decides whether a formula of the logic $\mathcal{PMDL}$ is satisfiable in a measurable structure from $\mathcal{PMDL}^{Meas}$ which runs in polynomial space.*

Proof. Let us first consider the formulas from MDL. Recall that in the proof of Theorem 2 we have shown that each $\theta \in \mathcal{L}_{MDL}$ is satisfiable in an MDL model iff it is satisfiable under our semantics. Thus we can use the result that there is a procedure for deciding whether a formula $\theta \in \mathcal{L}_{MDL}$ is satisfiable, that runs in polynomial space [13].

For probabilistic formulas, we want to use some parts of the proof of Theorem 2 (which in turn uses the proof of Theorem 1). Here we will use all the notations introduced in the proofs: Theorem 1 and Theorem 2. First, note that in the proof of Theorem 2 we proved, using filtration, that if a formula f is satisfiable, then it is satisfied in a model $M' = \langle s_2, \mathscr{X}', \mu', \tau' \rangle$ with m states, where m at most $2^{|Sub_{DL}(f)|}$, i.e., $s_2 = \{s_1, \ldots, s_m\}$, and where each state $s_i \in s_2$ is represented by its characteristic formula $\delta_i \in \Delta$. Now we will show that we can reduce the size of the set of states even more. Let $DS(f)$ denote the set of all deontic formulas ϕ such that $w(\phi)$ is a term that appears in the formula f (i.e., $w(\phi)$ is a sub-expression of f), Let us consider the set of equations and inequalities over the variables $x_1, \ldots, x_m$:

$$x_1 + \cdots + x_m = 1, \tag{4}$$

$$x_1 \geq 0, \ x_2 \geq 0, \ \ldots \ x_m \geq 0, \tag{5}$$

and, for each $\phi \in DS(f)$, the equation

$$\sum_{\delta_i \in \Delta_\phi} x_i = \mu'(\|\phi\|_{M'}), \tag{6}$$

where $\Delta_\phi = \{\delta \in \Delta \mid \delta \vdash \phi\}$. Here we employ the result form linear algebra which states that if a system of k linear equations has a *non-negative* solution, then it has a non-negative solution where at most k values are different than zero [4]. Since the above system has one solution, namely

$$(x_1, \ldots, x_m) = (\mu'(\{\delta_1\}), , \ldots, \mu'(\{\delta_m\})),$$

then the system of equations (4) and (6) has a non-negative solution with at most $k(f) = |DS(f)| + 1$ values different than zero (note that when we calculate the number of equations, we ignore (5), since it simply states non-negativity, which is already assumed). Without any loss of generality, assume that this solution assigns the values $x_i = d_i$, where $d_i = 0$ for $i > k(f)$. Then we can define $M = \langle S, \mathscr{X}, \mu, \tau \rangle$, where $S = \{s_1, \ldots, s_{k(f)}\}$, τ is the restriction of τ' to the set $S \subseteq s_2$, and for every $s_i \in S$, $\mu(\{s_i\}) = d_i$. Obviously $M' \models f$ implies $M \models f$, so it is sufficient to consider the structures with $k(f)$ worlds.

Now we describe our procedure which runs as follows: it systematically cycles through sets of characteristic formulas $\overline{\Delta} \subseteq \Delta$ of cardinality $k(f)$. Fixing such subsets can be obtained in polynomial space. Indeed, recall that each element of Δ is a conjunction of elements of $Sub_{DL}(f)$ and their negations, and the satisfiability of each conjunction in MDL can be checked in polynomial space [13]. Then, for each such $\overline{\Delta}$, we check if we can assign the probability values $x_1, \ldots, x_{k(f)}$ to its elements such that f is satisfied. We consider the formula which is the conjunction of the following formulas:

$$x_1 + \cdots + x_{k(f)} = 1,$$

$$x_1 \geq 0,\ x_2 \geq 0,\ \ldots\ ,\ x_{k(f)} \geq 0,$$

and the formula

$$Trans_{RCF}(f),$$

where $Trans_{RCF}(f)$ is obtained from f by applying the following transformations:

- we replace in f each occurrence of every $w(\phi)$ (for every $\phi \in DS(f)$) with

$$\sum_{\delta_i \in \Delta_\phi \cap \overline{\Delta}} x_i.$$

- We rewrite every integer coefficient from f with an expression that uses only 1, 0, and -1, using the binary representation of the numbers, and the powers are represented using multiplication. For example, number 9 is rewritten as $(1+1)(1+1)(1+1)+1$.

In this way, the size of obtained conjunction stays polynomial wrt. length of f. With this transformation, we directly follow the approach of [7]. The idea is that the obtained formula is a quantifier-free formula in the language of real closed fields (RCF). Then Canny's procedure [2], which decides satisfiability of quantifier-free formulas of RCF in polynomial space, can be applied. It is clear that f is satisfiable in $\mathcal{PMDL}^{Meas}$ iff for some $\overline{\Delta}$ the formula above is satisfied in RCF. This completes our proof. $\square$

6 The logic $\mathcal{PMDL}^2$

In this section, we present the logic $\mathcal{PMDL}^2$ whose language extends the language of $\mathcal{PMDL}$. This new logic assumes a fixed finite set of agents Ag, and it allows nesting of probabilities, enabling formulas that can express the uncertainty of one agent about some other agent's uncertainty about norms. Consequently, the logic $\mathcal{PMDL}^2$ will have a different probability structure, compared to the previous logic. Instead of having one measure μ over the states, we will have a function $\mathscr{P}$ that assigns a probability space to each agent and state ranging over a subset of all states. In the following sections, we will introduce the changes made to $\mathcal{PMDL}$ in order to construct $\mathcal{PMDL}^2$.

6.1 Syntax and Semantics

Definition 11 (Formulae). *Let $\mathbb{P}$ be a set of atomic propositions, and let Ag be a set of agents. The language $\mathcal{L}_{\mathcal{PMDL}^2}$ is generated by the following two sentences of BNF:*

$$[\mathcal{L}_{MDL}] \quad \phi ::= p \mid \neg\phi \mid \phi \wedge \phi \mid O\phi \qquad\qquad p \in \mathbb{P}$$
$$[\mathcal{L}_{\mathcal{PMDL}^2}] \quad \theta ::= \phi \mid a_1 w_i(\theta_1) + \cdots + a_n w_i(\theta_n) \geq \alpha \mid \neg\theta \mid \theta \wedge \theta \quad a_j, \alpha \in \mathbb{N}, i \in Ag$$

The expression $w_i(\phi) \geq \alpha$ stands for "according to the agent i, the probability of ϕ is at least α".

Note that the formula $a_1 w_i(\theta_1) + \cdots + a_n w_i(\theta_n) \geq \alpha$ contains exclusively one agent i; such a formula is called i−probability formula. Although we do not allow combination of agents within one linear combination, the formulas within the scope of w_i might contain probabilities of other agents than i, as illustrated by the following example.

Example 2. *Following our previous example about behavioral norms in a library, we can now express the certainty of a person about another person's certainty. For example, the fact that a person has become fairly certain that another person is certain about it not being normal to be quiet in a library. This might be expressed by the probabilistic statement "agent i attributes the probability that agent j attributes the probability that one is obliged to be quiet to be at most 0.2 is at least 0.9". This sentence could be formalized using the introduced language as*

$$w_i(w_j(Oq) \leq 0.2) \geq 0.9.$$

For the formulas of $\mathcal{PMDL}^2$, we introduce the same types of abbreviations as we have done for $\mathcal{PMDL}$.

Now we introduce the semantics of $\mathcal{PMDL}^2$.

Probabilistic Deontic Logic

Definition 12 (Model). *A $\mathcal{PMDL}^2$ model is a tuple $M = \langle S, \tau, \mathscr{P} \rangle$, where:*

- *S is a non-empty set of states*

- *τ associates with each state $s \in S$ a tuple containing an MDL model and one of its worlds: $\tau(s) = (D_s, w_s)$ where:*
  - *$D_s = (W_s, R_s, V_s) \in \mathbb{D}$*
  - *$w_s \in W_s$*

- *$\mathscr{P}(i, s)$ is a function assigning to each combination of agent (i) and state (s) a probability space $\mathscr{P}(i, s) = (S_{i,s}, \mathscr{X}_{i,s}, \mu_{i,s})$ where:*
  - *$S_{i,s} \subseteq S$ an arbitrary subset of S that can be interpreted as the set of states that agent i has conceptions about in state s.*
  - *$\mathscr{X}_{i,s}$ is a σ-algebra of subsets of $S_{i,s}$*
  - *$\mu_{i,s} : \mathscr{X}_{i,s} \mapsto [0, 1]$ is a probability measure.*

Let us illustrate this definition.

Example 2. *(continued)* Assume a finite set of atomic propositions $\{p, q\}$. Let us consider the model $M = \langle S, \tau, \mathscr{P} \rangle$, where

- $S = \{s_1, s_2, s_3, s_4\}$

- $\mathscr{P}$ We will set the probability measures explicitly for each state-agent pair while the respective set $S_{i,s}$ will be set to S and the respective sigma-algebra $\mathscr{X}_{i,s}$ will be the power set of S.

  - μ_{i,s_1} is characterized by: $\mu_{i,s_1}(\{s_1\}) = 0.5$, $\mu_{i,s_1}(\{s_2\}) = \mu_{i,s_1}(\{s_3\}) = 0.2$, $\mu_{i,s_1}(\{s_4\}) = 0.1$
  - μ_{j,s_1} is characterized by: $\mu_{j,s_1}(\{s_1\}) = \mu_{j,s_1}(\{s_2\}) = 0.1$, $\mu_{j,s_1}(\{s_3\}) = 0.0$, $\mu_{j,s_1}(\{s_4\}) = 0.8$.
  - μ_{j,s_2} is characterized by: $\mu_{j,s_2}(\{s_1\}) = \mu_{j,s_2}(\{s_2\}) = 0.0$, $\mu_{j,s_2}(\{s_3\}) = 0.1$, $\mu_{j,s_2}(\{s_4\}) = 0.9$.
  - μ_{j,s_3} is characterized by: $\mu_{j,s_3}(\{s_1\}) = \mu_{j,s_3}(\{s_2\}) = \mu_{j,s_3}(\{s_3\}) = 0.0$, $\mu_{j,s_3}(\{s_4\}) = 1$.
  - μ_{j,s_4} is characterized by: $\mu_{j,s_4}(\{s_1\}) = 0.5$, $\mu_{j,s_4}(\{s_2\}) = \mu_{j,s_4}(\{s_3\}) = 0.1$, $\mu_{j,s_4}(\{s_4\}) = 0.3$.

- τ maps each state in S to a pointed deontic model; specifically for our interest is the assignment of state s_1, $D_{s_1} = (W_{s_1}, R_{s_1}, V_{s_1})$ and w_{s_1} such that

  - $W_{s_1} = \{w_1, w_2, w_3, w_4\}$
  - $R_{s_1} = \{(w_1, w_2), (w_1, w_3), (w_2, w_2), (w_2, w_3), (w_3, w_2), (w_3, w_3), (w_4, w_2),$ $(w_4, w_3), (w_4, w_4)\}$
  - $V_{s_1}(p) = \{w_1, w_3\}$, $V_{s_1}(q) = \{w_2, w_3\}$
  - $w_{s_1} = w_1$

Note that the domain of τ is always the whole set S; but in this example, we only explicitly specify $\tau(s_1)$ for illustration purposes.

This model is depicted in Figure 2. The circle on the right contains the four states of the model. The dotted lines represent probability measure μ_{i,s_1} the others are not drawn to reduce cluttering. Each of the states is equipped, by τ, with a standard pointed model of MDL. In this picture, only one of them is shown, the one that corresponds to s_1. It is represented within the circle on the left. Note that the arrows depict the relation R. If we assume that q stands for "quiet", like in the previous example, in all R–successors of w_1 the proposition q holds. Note that, according to Definition 3, this means that in w_1 people are obliged to be quiet in the library.

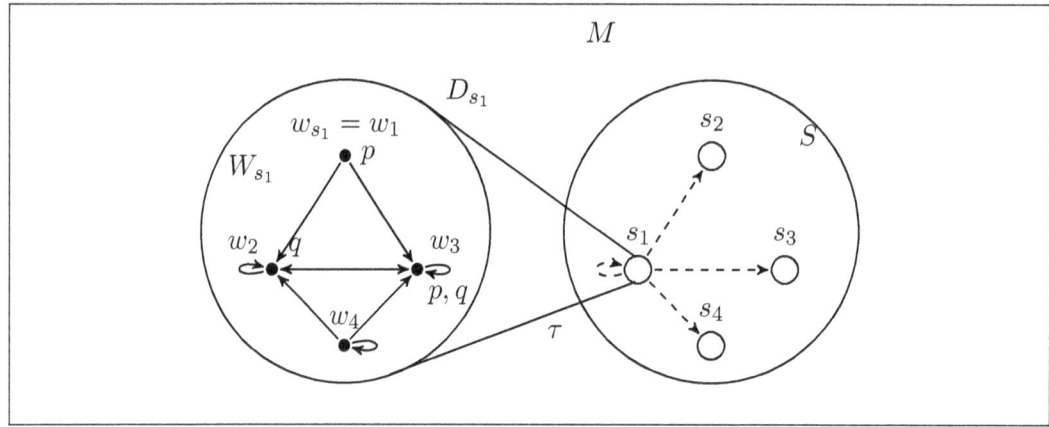

Figure 2: Model $M = \langle S, \tau, \mathscr{P} \rangle$.

Next, the satisfiability of a formula in a model can be defined. First, the truth of a deontic formula in a state of a $\mathcal{PMDL}^2$ model is given. This definition is in accordance with the standard satisfiability relation of MDL $\models_{MDL}$.

Definition 13 (Satisfaction). *Let* $M = \langle S, \tau, \mathscr{P} \rangle$ *be a* $\mathcal{PMDL}^2$ *model, and let* $s \in S$. *We define the satisfiability of formula* $\theta \in \mathcal{L}_{\mathcal{PMDL}^2}$, *in state s of model M denoted by $M, s \models \theta$ recursively as follows with* $\phi \in \mathcal{L}_{MDL}$:

- $M, s \models \phi$ iff $D_s, w_s \models_{MDL} \phi$, where $\tau(s) = (D_s, w_s)$.
- $M, s \models a_1 w_i(\theta_1) + \cdots + a_n w_i(\theta_n) \geq \alpha$ iff
  $a_1 \mu_{i,s}(\|\theta_1\|_{i,s}^M) + \cdots + a_n \mu_{i,s}(\|\theta_n\|_{i,s}^M) \geq \alpha$.
- $M, s \models \neg \theta$ iff $M, s \not\models \theta$.
- $M, s \models \theta_l \wedge \theta_k$ iff $M, s \models \theta_l$ and $M, s \models \theta_k$.

For a model $M = \langle S, \tau, \mathscr{P} \rangle$, *a formula* $\theta \in \mathcal{L}_{\mathcal{PMDL}^2}$, *state s and agent i, let* $\|\theta\|_{i,s}^M$ *denote the set of states that satisfy θ, from the perspective of agent i in state s i.e.,*

$$\|\theta\|_{i,s}^M = \{s_2 \in S_{i,s} \mid M, s_2 \models \theta\}.$$

We omit the super- and subscripts from $\|\theta\|_{i,s}^M$ *when it is clear from context. The satisfaction relation shows that in this model construction formulas θ can occur as the argument to a weight formula w_i, this means that weight formulas can be arguments of weight operators.*

Since the focus is on measurable structures and completeness is proven for this class of structures, this class is redefined for $\mathcal{PMDL}^2$ models.

Definition 14 (Measurable model). *A probabilistic deontic model is* measurable *if*

$$\|\phi\|_{i,s}^M \in \mathscr{X}_{i,s}$$

for every $\phi \in \mathcal{L}_{MDL}$.

Example 2. *(continued)* Continuing the previous example, according to Definition 13 it holds that $M, s_1 \models Oq$. At this point it is also possible to speak of the uncertainty of agent i about the uncertainty of agent j of the obligation to be quiet in the library. Assume that τ is defined in the way such that $M, s_2 \models Oq$ and $M, s_3 \models Oq$, but $M, s_4 \not\models Oq$. Then $\mu_{j,s_1}(\|Oq\|) = \mu_{j,s_1}(\{s_1, s_2, s_3\}) = 0.1 + 0.1 + 0.0 = 0.2$; $\mu_{j,s_2}(\|Oq\|) = \mu_{j,s_2}(\{s, s_2, s_3\}) = 0.0 + 0.0 + 0.1 = 0.1$; $\mu_{j,s_3}(\|Oq\|) = \mu_{j,s_3}(\{s, s_2, s_3\}) = 0.0 + 0.0 + 0.0 = 0.0$; $\mu_{j,s_4}(\|Oq\|) = \mu_{j,s_4}(\{s_1, s_2, s_3\}) = 0.5 + 0.1 + 0.1 = 0.7$. From this follows that $\mu_{i,s_1}(\|w_j(Oq) \leq 0.2\|) = \mu_{i,s_1}(\{s_1, s_2, s_3\}) = 0.5 + 0.2 + 0.2 = 0.9$. According to Definition 6, $M, s_1 \models w_i(w_j(Oq) \leq 0.2) \geq 0.9$. Describing the uncertainty of agent i about the uncertainty of agent j's obligation to be quiet in the library.

6.2 Axiomatization

The following axiomatization $AX_{\mathcal{PMDL}^2}$ combines –like $AX_{\mathcal{PMDL}}$– the axioms of proof system D of MDL [14] with the axioms of the probabilistic logic. In this case, the probabilistic axioms come from [6].

The Axiomatic System: $AX_{\mathcal{PMDL}^2}$

Tautologies and Modus Ponens

Taut. All instances of propositional tautologies.

MP. From θ and $\theta \to \theta'$ infer θ'.

Reasoning with O:

O-... see axiomatization in Section 3

Reasoning about linear inequalities:

I1.-I7. see axiomatization in Section 3

Reasoning about probabilities:

W1. $w_i(\theta) \geq 0$ (non negativity).

W2. $w_i(\theta \vee \theta') = w_i(\theta) + w_i(\theta')$, if $\neg(\theta' \wedge \theta')$ is an instance of a classical propositional tautology (finite additivity).

W3. $w_i(\top) = 1$

P-Dis. From $\theta \leftrightarrow \theta'$ infer $w_i(\theta) = w_i(\theta')$ (probabilistic distributivity)

As before the axiom Taut allows all propositional tautologies. Though since $\mathcal{L}_{MDL}$ is included in $\mathcal{L}_{\mathcal{PMDL}^2}$ the distinction for Modus Ponens (MP) dissolves and can be applied to both types of formulas. P-Dis is an inference rule which states that two equivalent deontic formulas must have the same probability values.

6.3 Soundness and Completeness

In this section it is proven that the construction $\mathcal{PMDL}^2$ is sound and complete with respect to the class of measurable models, combining and adapting the approaches from [14, 6].

Theorem 4 (Soundness & Completeness). *The axiom system $AX_{\mathcal{PMDL}^2}$ is sound and complete with respect to the class of measurable probabilistic deontic models. i.e., $\vdash \theta$ iff $\models \theta$.*

Proof. The proof is a modification of the corresponding proof for $\mathcal{PMDL}$. To prove completeness, we need to show that every consistent formula θ is satisfiable in a measurable model. The modification of the logic gives iterations of weight formulas of arbitrary depth, also instead of one measure there is a measure for each agent and state pair (i, s); for this, the proof needs to be adjusted.

For any formula θ we will denote the set of sub-formulas closed under negation as follows $Sub^+(\theta) = Sub(\theta) \cup \{\neg\theta' \mid \theta' \in Sub(\theta)\}$. We say that a set of formulas $A \subseteq B$ is maximal with regards to B when $\forall \theta \in B$, A contains either θ or $\neg\theta$.

Let θ be a consistent formula of $\mathcal{L}_{\mathcal{PMDL}^2}$. Then let S denote the set of maximal consistent subsets of $Sub^+(\theta)$. And define for each $s \in S$ the element $\xi_s = \bigwedge_{\theta' \in s} \theta'$ to be the conjunction of elements in s. Denote the set of elements ξ_s as follows $\Xi = \{\xi_s \mid s \in S\}$. S will be the set of states of our model of the formula θ. Furthermore, in order to define τ we construct for each state $s \in S$ the MDL context as $s_{MDL} = \{\phi \in \mathcal{L}_{MDL} \mid \phi \in s\}$ and its conjunction as $\delta_s = \bigwedge_{\phi \in s_{MDL}} \phi$. Then we can define τ in the following way. By completeness of MDL, for each δ_s there is a deontic model D_s and a world w_s in it such that $D_s, w_s \models_{MDL} \delta_s$. Then we define $\tau(s) = (D_s, w_s)$.

Since our probabilistic deontic model is of the form $M = (S, \tau, \mathscr{P})$ this leaves the task of defining the probability assignment $\mathscr{P}$. We chose $S_{i,s} = S$ and always assume that every subset of S is measurable. The rest of the proof is essentially the same as the corresponding proof for defining probability assignment for probabilistic epistemic logic from [6]. $\mathscr{P}$ has to be defined in such a way that when we consider the model M, then for every $s \in S$ and every formula $\psi \in Sub^+(\theta)$ we have $M, s \models \psi$ iff $\psi \in s$. To do this we will make use of additivity using the following equivalence:

$$\vdash \psi \leftrightarrow \bigvee_{\{s \in S \mid \psi \in s\}} \xi_s.$$

Then using the axiom system, for every $i \in Ag$ we can derive in a similar way as in proof of Theorem 1 the following equation:

$$\vdash w_i(\psi) = \sum_{\{s \in S \mid \psi \in s\}} w_i(\xi_s).$$

By I1-I7, this can be extended to show that any i-probability formula $\theta' \in Sub^+(\theta)$ can be equivalently rewritten as a formula of the form $\sum_{s_2 \in S} c_{s_2} \mu_{i,s}(s_2) \geq \alpha$.

Similarly as in the proof of Theorem 1 we encode the problem to a set of linear equations and inequalities over variables of the form x_{iss_2}, where x_{iss_2} represents $\mu_{i,s}(\{s_2\})$. For each i-probability formula $\psi \in Sub^+(\theta)$ we have a corresponding inequality. Using the conclusion et the end of the previous paragraph, When $\psi \in s$ then the corresponding inequality is: $\sum_{s_2 \in S_{i,s}} c_{s_2} x_{iss_2} \geq \alpha$. When $\neg \psi \in s$, then we have $\sum_{s_2 \in S_{i,s}} c_{s_2} x_{iss_2} < \alpha$. We also add the non-negativity constraints, and the condition that $\sum_{s_2 \in S_{i,s}} x_{iss_2} = 1$, as in the proof of Theorem 1. Furthermore, following that proof and [7], one can show that this system of inequalities has a solution $x^*_{iss_2}$ for all $s_2 \in S_{i,s}$; since each ξ_s is consistent. The solution of this large system determines probability values of each agent in each state.

What is left to show is that for every formula $\psi \in Sub^+(\theta)$ and every state in S, we have $M, s \models \psi$ iff $\psi \in s$. The proof proceeds by induction on ψ. If ψ is a deontic formula the result is immediate from the definition of τ. The cases where ψ is a negation or conjunction are straightforward. The case where ψ is an i-probability formula follows from the construction above. Therefore if the formula θ is consistent then it is satisfiable in a model. $\square$

6.4 Decidability

Finally, we show that the logic $\mathcal{PMDL}^2$ is decidable.

Theorem 5. *Satisfiability problem for $\mathcal{PMDL}^2$ is decidable.*

Proof. Similarly, as in the proof of Theorem 2, we combine the method of filtration and reduction to finite systems of inequalities. Because of the similarity, we omit some details. In the proof, we will use some notation already introduced in the paper. Let us assume that the formula θ has a model $M = \langle S, \tau, \mathscr{P} \rangle$, where $\mathscr{P}(i, s) = (S_{i,s}, \mathscr{X}_{i,s}, \mu_{i,s})$. We will use filtration to construct the model of θ with finitely many states. By $\sim$ we denote the equivalence relation over $S \times S$, where $s \sim s_2$ iff for every $\theta' \in Sub(\theta)$, $s \models \theta'$ iff $s_2 \models \theta'$. Then the quotient set $S_{/\sim}$ is of the size $|S_{/\sim}| \leq 2^{|Sub(\theta)|}$. As before, for every class C_j we choose an element and denote it s_j. We consider the model $M^* = \langle S^*, \tau^*, \mathscr{P}^* \rangle$, in which:

- $S^* = \{s_j \mid C_j \in S_{/\sim}\}$,
- $\mathscr{P}^*(i, s_j) = (S^*_{i,s_j}, \mathscr{X}^*_{i,s_j}, \mu^*_{i,s_j})$ such that:
  - $S^*(i, \varepsilon_j) = \{s_{l_0} \in S^* \mid (\exists s_2 \in C_{s_k}) s_2 \in S(s_j)\}$,
  - $\mathscr{X}^*_{i,s_j}$ is the power set of $S^*(i, s_j)$,

- $\mu_{i,s_j}^*(\{s_k\}) = \mu_{i,s_j}(C_{w_k})$ (and μ_{i,s_j}^* extends to $\mathscr{X}_{i,s_j}^*$ by additivity),

- $\tau^*(s_j) = \tau(s_j)$.

It can be shown that M^* is a measurable model. Moreover, using straightforward induction on the complexity of the formula, one can show that for any $\theta' \in Sub(\theta)$, $M, s \models \theta'$ iff $M^*, s_i \models \theta'$ where s_i represents C_s in M^*. Additionally in the same way, as in the proof of Theorem 2, we can show that the number of worlds in a deontic model of each state is finite and at most exponential wrt. size of $Sub(\theta)$. As the number of propositional letters and agents from θ is also finite, it turns out that we have to check only finitely many options for the choice of S and τ.

Let us describe the procedure which checks the satisfiability of a formula θ. First, we transform θ to a disjunction of the formulas of the form $\bigwedge_{k=1}^{|Sub(\theta)|} \psi_k$, where $\psi_k \in Sub^+(\theta)$ and each subformula of θ appears exactly once in each conjunction (either negated or not). The conjunctions whose sub-conjunction consisting of deontic formulas is unsatisfiable can be eliminated using the decidability of MDL, as we have done in the proof of decidability of $\mathcal{PMDL}$. In each state $s \in S^*$ exactly one formula of the form $\bigwedge_{k=1}^{|Sub(\theta)|} \psi_k$ holds. Denote that (characteristic) formula by δ_s as before. Here we slightly abuse the notation, and we write $\psi \in \delta_s$ if ψ is a conjunct in δ_s. For every $\ell \leq 2^{Sub(\theta)}$ we will consider ℓ formulas of the above form such that the following three conditions hold:

- Those formulas δ_s are not necessarily different, but each formula does not contain both ψ and $\neg \psi$ in the top conjunction.

- The conjunction of all deontic formulas from the top conjunction is consistent.

- At least one δ_s must contain θ in the top conjunction.

Then for every agent i, every state s_j, $j < \ell$, we consider the following set of equations and inequalities, with the set of variables x_{iss_2}, where x_{iss_2} represents $\mu_{i,s}(\{s_2\})$ (as in the proof of completeness).

$$\sum_{s_2} x_{iss_2} = 1,$$

$$x_{iss_2} \geq 0,$$

$$a_1 \sum_{s_k : \theta_1 \in \delta_{s_k}} x_{is_j s_k} + \cdots + a_n \sum_{s_k : \theta_n \in \delta_{s_k}} x_{is_j s_k} \geq \alpha,$$

whenever $(a_1 w_i(\theta_1) + \cdots + a_n w_i(\theta_n) \geq \alpha) \in \delta_{s_j}$,

$$a_1 \sum_{s_k : \theta_1 \in \delta_{s_k}} x_{i s_j s_k} + \cdots + a_n \sum_{s_k : \theta_n \in \delta_{s_k}} x_{i s_j s_k} < \alpha,$$

whenever $\neg(a_1 w_i(\theta_1) + \cdots + a_n w_i(\theta_n) \geq \alpha) \in \delta_{s_j}$.

Thus we have translated the problem of satisfiability of θ to a decidable problem of solving systems of linear inequalities, as before. Since we have finitely many possibilities for the choice of ℓ, and for each ℓ finitely many possibilities to choose ℓ characteristic formulas, our logic $\mathcal{PMDL}^2$ is decidable. □

7 Conclusion

In this article, we introduced two probabilistic deontic logics. Each of them extends both monadic deontic logic and probability logic from [7]. The language of the first logic, $\mathcal{PMDL}$ is designed for reasoning about the probability of deontic statements. We axiomatized that language and proved soundness and completeness with respect to corresponding semantics. We also proved that our logic is decidable in PSPACE. The second proposed language allows nested probability operators, and it allows to express the uncertainty of one agent about the uncertainty that another agent places on deontic statements.

To the best of our knowledge, we are the first to propose logical frameworks of probabilistic deontic logics for reasoning with uncertainty about norms. It is worth mentioning that there is a recent knowledge representation framework about probabilistic uncertainty in deontic reasoning obtained by merging deontic argumentation and probabilistic argumentation frameworks [15].

Our logic $\mathcal{PMDL}$ used MDL as the underlying framework, we used this logic simply because it is one of the most studied deontic logics. On the other hand, MDL is also criticized because of some issues [9], like the representation of contrary-to-duty obligations. It is important to point out that the axiomatization technique developed in this work can also be applied if we replace MDL with, for example, dyadic deontic logic, simply by changing the set of deontic axioms and the function τ in the definition of the model, which would lead to a more expressive framework for reasoning about uncertain norms. Another avenue for future research is to extend the language by allowing conditional probabilities. In such a logic, it would be possible to express that one uncertain norm becomes more certain if another norm is accepted or learned.

References

[1] Boella, G., van der Torre, L.W.N., Verhagen, H.: Introduction to normative multiagent systems. Comput. Math. Organ. Theory **12**(2-3), 71–79 (2006), https://doi.org/10.1007/s10588-006-9537-7

[2] Canny, J.F.: Some algebraic and geometric computations in PSPACE. In: Proceedings of the 20th Annual ACM Symposium on Theory of Computing, May 2-4, 1988, Chicago, Illinois, USA. pp. 460–467 (1988)

[3] Chellas, B.F.: Modal Logic: An Introduction. Cambridge University Press (1980). https://doi.org/10.1017/CBO9780511621192

[4] Chvátal, V.: Linear Programming. Series of books in the mathematical sciences, W.H. Freeman (1983), https://books.google.pt/books?id=JkHBQgAACAAJ

[5] Dong, H., Liao, B., Markovich, R., van der Torre, L.W.N.: From classical to non-monotonic deontic logic using aspic$^+$. In: Blackburn, P., Lorini, E., Guo, M. (eds.) Logic, Rationality, and Interaction - 7th International Workshop, LORI 2019, Chongqing, China, October 18-21, 2019, Proceedings. Lecture Notes in Computer Science, vol. 11813, pp. 71–85. Springer (2019). https://doi.org/10.1007/978-3-662-60292-8_6, https://doi.org/10.1007/978-3-662-60292-8_6

[6] Fagin, R., Halpern, J.Y.: Reasoning about knowledge and probability. J. ACM **41**(2), 340–367 (Mar 1994), https://doi.org/10.1145/174652.174658

[7] Fagin, R., Halpern, J.Y., Megiddo, N.: A logic for reasoning about probabilities. Information and Computation **87**(1), 78 – 128 (1990). https://doi.org/https://doi.org/10.1016/0890-5401(90)90060-U, http://www.sciencedirect.com/science/article/pii/089054019090060U, special Issue: Selections from 1988 IEEE Symposium on Logic in Computer Science

[8] Frisch, A., Haddawy, P.: Anytime deduction for probabilistic logic. Artificial Intelligence **69**, 93–122 (1994)

[9] Hansen, J.: The paradoxes of deontic logic: Alive and kicking. Theoria **72**, 221 – 232 (09 2006). https://doi.org/10.1111/j.1755-2567.2006.tb00958.x

[10] van der Hoek, W.: Some considerations on the logic pfd~. J. Appl. Non Class. Logics **7**(3) (1997)

[11] Horty, J.F.: Agency and Deontic Logic. Oxford University Press (2001)

[12] Hughes, G.E., Cresswell, M.J.: A companion to modal logic. Methuen London; New York (1984)

[13] Ladner, R.E.: The computational complexity of provability in systems of modal propositional logic. SIAM J. Comput. **6**(3), 467–480 (1977). https://doi.org/10.1137/0206033, https://doi.org/10.1137/0206033

[14] Parent, X., Van Der Torre, L.: Introduction to Deontic Logic and Normative Systems. Texts in logic and reasoning, College Publications (2018), https://books.google.nl/books?id=IyUYwQEACAAJ

[15] Riveret, R., Oren, N., Sartor, G.: A probabilistic deontic argumentation framework. International Journal of Approximate Reasoning **126**, 249 –

271 (2020). https://doi.org/https://doi.org/10.1016/j.ijar.2020.08.012, http://www.sciencedirect.com/science/article/pii/S0888613X20302188

[16] Sarathy, V., Scheutz, M., Malle, B.F.: Learning behavioral norms in uncertain and changing contexts. In: 2017 8th IEEE International Conference on Cognitive Infocommunications (CogInfoCom). pp. 000301–000306 (2017). https://doi.org/10.1109/CogInfoCom.2017.8268261

[17] Savic, N., Doder, D., Ognjanovic, Z.: Logics with lower and upper probability operators. Int. J. Approx. Reason. **88**, 148–168 (2017). https://doi.org/10.1016/j.ijar.2017.05.013, https://doi.org/10.1016/j.ijar.2017.05.013

[18] Tomic, S., Pecora, F., Saffiotti, A.: Learning normative behaviors through abstraction. In: Giacomo, G.D., Catalá, A., Dilkina, B., Milano, M., Barro, S., Bugarín, A., Lang, J. (eds.) ECAI 2020 - 24th European Conference on Artificial Intelligence, 29 August-8 September 2020, Santiago de Compostela, Spain, August 29 - September 8, 2020 - Including 10th Conference on Prestigious Applications of Artificial Intelligence (PAIS 2020). Frontiers in Artificial Intelligence and Applications, vol. 325, pp. 1547–1554. IOS Press (2020), https://doi.org/10.3233/FAIA200263

[19] Tomovic, S., Ognjanovic, Z., Doder, D.: A first-order logic for reasoning about knowledge and probability. ACM Trans. Comput. Log. **21**(2), 16:1–16:30 (2020)

[20] de Wit, V., Doder, D., Meyer, J.C.: A probabilistic deontic logic. In: ECSQARU. Lecture Notes in Computer Science, vol. 12897, pp. 616–628. Springer (2021)

[21] von Wrigth, G.H.: I. Deontic Logic. Mind **LX**(237), 1–15 (01 1951), https://doi.org/10.1093/mind/LX.237.1

Activation-Based Conditional Inference

Marco Wilhelm, Diana Howey, and Gabriele Kern-Isberner
Department of Computer Science, TU Dortmund University, Dortmund, Germany
`marco.wilhelm@tu-dortmund.de, diana.howey@tu-dortmund.de,`
`gabriele.kern-isberner@cs.tu-dortmund.de`

Kai Sauerwald and Christoph Beierle
Faculty of Mathematics and Computer Science, FernUniversität in Hagen, Hagen, Germany
`kai.sauerwald@fernuni-hagen.de, christoph.beierle@fernuni-hagen.de`

Abstract

Activation-based conditional inference (ActInf) combines conditional reasoning and ACT-R, a cognitive architecture developed to formalize human reasoning, and therewith provides a powerful inference formalism which makes it possible to integrate several aspects of human reasoning, such as focusing, forgetting, and remembering, into formal uncertain reasoning. The basic idea of activation-based conditional inference is to determine a reasonable, cognitively adequate subset of a conditional belief base before drawing inductive inferences. Central to activation-based conditional inference is the activation function which assigns to the conditionals in the belief base a degree of activation mainly based on the conditional's relevance for the current query and its usage history. Here, we develop a blueprint for activation-based conditional inference and illustrate how focusing, forgetting, and remembering are included within our framework.

1 Introduction

Knowledge-based systems, also called *expert systems* [9, 18], are computer programs which help to infer information from expert knowledge in order to solve complex reasoning tasks from the especially nowadays fast growing amount of knowledge. Typically, knowledge-based systems consist of two components, a domain-specific belief base and an inference engine. When a user inserts a query, the belief base and the query are transferred to the inference engine which decides whether the query can be answered from the belief base or not. To make a meaningful decision, a sophisticated inference methodology has to be implemented in the inference engine. Taken as a whole, the aspiration of expert systems is to draw inferences of high quality from usually incomplete and uncertain beliefs. The contribution of *activation-based conditional inference (ActInf)* presented in this paper to this inference process is a preselection of relevant beliefs from the belief base before this so-reduced belief base is transferred to the inference engine with the objective to both (a) reduce computational costs during the inference process and (b) model human cognitive processes within expert systems more adequately.

It is obvious and reasonable that human reasoners do not draw inferences based on all of their beliefs, in particular when they have to make their decisions in due time. Basically, there are two cognitive processes which affect the selection of beliefs: The long-term process of forgetting and remembering and the short-term process of activating specific beliefs depending on the context. In *ACT-R (Adaptive Control of Thought-Rational,* [5, 4]), a well-founded cognitive architecture established in cognitive science with the aim to formalize human reasoning, the long-term memory is represented by the *base-level activation*, while the context-dependent activation of beliefs is described by the *spreading activation theory* [3]. The core idea behind the spreading activation theory is that an initial priming caused by sensory stimuli triggers certain *cognitive units* [3] which again trigger related cognitive units and so on until the disposition for activation is too low to trigger further cognitive units. The triggered cognitive units settle the current focus in which reasoning takes place. However, while ACT-R provides in general a very interesting and intuitive cognitive environment for reasoning processes, it is severely limited from the perspective of knowledge representation because it is based crucially on using production rules as reasoning engine.

In this paper, we reinterpret the basic elements of ACT-R in a more abstract form, detaching them from its inference engine using production rules. In this way, we make ACT-R more broadly usable for knowledge representation. Then we combine these abstract ACT-R elements with state-of-the-art conditional reasoning operators, presenting activation-based conditional inference as a novel approach to reasoning that explicitly takes cognitive aspects into account. Activation-based conditional inference adapts the concept of (de)activation of knowledge entities from ACT-R and combines it with conditional inference formalisms from nonmonotonic reasoning. Here, we define a model for activation-based conditional inference by applying the activation function from ACT-R, constituting of the base-level activation and the spreading activation, to conditional statements of the form $(B|A)$ with the meaning "if A holds, then usually B holds, too." In this way, on the one hand, we generalize the concept of *focused inference* from [23] which involves a selection of conditionals from a belief base based on their syntactical linkage and give it a profound cognitive meaning. On the other hand, we equip ACT-R, which is typically realized as a *production system* [12, 16], with a modern inference formalism of high quality.

The activation of conditionals and the drawing of inferences from the reduced belief base are two separate processes. The modular structure of activation-based conditional inference allows the user to exchange both the actual configuration of the activation function and the inference formalism independently. For the activation function we give a blueprint which is motivated by the cognitive theory behind ACT-R and which behaves well in all of our examined examples. As possible inference formalisms, we consider System P [1, 13], System Z [17], and c-representations [10], each of them showing characteristic properties which are beneficial in the context of activation-based conditional inference (in particular, *Semi-Monotony*, *Maximal Normality*, and, *Syntax Splitting*, respectively). In [22], we have applied activation-based conditional inference to *probabilistic* conditionals $(B|A)[x]$, meaning that "if A holds, then B follows with probability x," which shows that our approach also works in the quantitative setting.

This paper is a revised and largely extended version of [21] and is organized as follows. First, we recall some basics on conditional logics, inductive inference formalisms, and focused inference. Then, we briefly discuss the ACT-R architecture while highlighting the activation function as the basis of the selection strategy for retrieving knowledge entities in ACT-R. Afterwards, we give an outline of our activation-based conditional inference approach and examine the developed activation function for conditionals in detail. Finally, we show how the concepts of forgetting and remembering can be integrated into our framework before we conclude with a summary and an outlook.

2 Logical Foundations

In this section we recall the logical foundations of qualitative conditional reasoning. We introduce a propositional logic $\mathcal{L}(\Sigma)$ as a background theory for conditionals, extend this logic to the logic of conditionals $\mathcal{CL}(\Sigma)$, and discuss some prominent inductive inference formalisms which can be used to draw inferences from (finite) sets of conditionals $\Delta \subseteq \mathcal{CL}(\Sigma)$. Finally, we recall the idea of *focused inference* from [23] which constitutes the connecting factor for our investigations on *activation-based conditional inference*.

2.1 Propositional Logic

We consider a *propositional language* $\mathcal{L}(\Sigma)$ defined over a finite set of *propositional variables* Σ. Elements in Σ are denoted by lowercase letters $(a, b, c, \ldots)$ and are called *atoms* for short. *Formulas* in $\mathcal{L}(\Sigma)$ (also called *propositions* and denoted by uppercase letters $A, B, C \ldots$) are either atoms from Σ or compounded formulas built by the common connectives $\neg$ (*negation*), $\wedge$ (*conjunction*), and $\vee$ (*disjunction*). A *literal* is either an atom or its negation.

The semantics of formulas in $\mathcal{L}(\Sigma)$ is given by *interpretations* $I : \Sigma \to \{0, 1\}$ which assign atoms $a \in \Sigma$ a *truth value*, either $I(a) = 1$ (a is *true* in the interpretation I) or $I(a) = 0$ (a is *false* in I). Interpretations are recursively extended to compounded formulas by $I(\neg A) = 1$ iff $I(A) = 0$, $I(A \wedge B) = 1$ iff $I(A) = 1$ and $I(B) = 1$, as well as $I(A \vee B) = 1$ iff $I(A) = 1$ or $I(B) = 1$, as usual in propositional logics. The set of all interpretations over Σ is denoted by $\mathcal{I}(\Sigma)$. An interpretation $I \in \mathcal{I}(\Sigma)$ is a *model* of a formula A iff $I(A) = 1$. A formula A from $\mathcal{L}(\Sigma)$ *entails* another formula B from $\mathcal{L}(\Sigma)$, written $A \models B$, iff every model $I \in \mathcal{I}(\Sigma)$ of A is also a model of B, i.e., iff $I(A) = 1$ implies $I(B) = 1$. Iff A entails B and B entails A, then A and B are *logically equivalent*, $A \equiv B$ in symbols.

In order to shorten expressions, we use the abbreviations $AB = A \wedge B$ (*juxtaposition*), $\overline{A} = \neg A$ (*overline*), $A \Rightarrow B = \overline{A} \vee B$ (*material implication*), $\top = A \vee \overline{A}$ (*tautology*), and $\bot = A\overline{A}$ (*contradiction*). In order to refer to the set of atoms which are mentioned in a formula A, we write $\Sigma(A)$. That is, $\Sigma(A)$ is the *signature* of A.

2.2 Conditionals and Ranking Semantics

To be able to formalize *uncertain beliefs* of a reasoner, we extend the propositional language $\mathcal{L}(\Sigma)$ by the use of the conditional operator $|$ and obtain the language of defeasible *(propositional) conditionals*

$$\mathcal{CL}(\Sigma) = \{(B|A) \mid A, B \in \mathcal{L}(\Sigma)\}.$$

Conditionals $(B|A) \in \mathcal{CL}(\Sigma)$ have the intuitive meaning "if A holds, then usually B holds, too," which means that B is a plausible consequence of A but there might be exceptional cases in which the conclusion from A to B fails. *Plausible facts* are subsumed within $\mathcal{CL}(\Sigma)$ by conditionals $(A|\top)$ stating that A is assumed to hold without any precondition. Finite sets of conditionals serve as *belief bases*.

The formal semantics of conditionals is based on *ranking functions* over *possible worlds*. The possible worlds considered here are simply the interpretations in $\mathcal{I}(\Sigma)$ represented as *complete conjunctions of literals*. That is, the possible world which refers to the interpretation $I \in \mathcal{I}(\Sigma)$ mentions the atoms $a \in \Sigma$ that are *true* in I

(i.e., $I(a) = 1$) positively, and the atoms $a \in \Sigma$ that are *false* in I (i.e., $I(a) = 0$) occur in the possible world in negated form ($\neg a$). The set of all possible worlds over Σ is denoted by $\Omega(\Sigma)$.

By following [20], *ranking functions* $\kappa : \Omega(\Sigma) \to \mathbb{N}_0^\infty$ map possible worlds to a degree of implausibility while satisfying the normalization condition $\kappa^{-1}(0) \neq \emptyset$. The higher the rank of a possible world, the less plausible the possible world is. Hence, $\kappa^{-1}(0)$ is the set of the most plausible worlds. Ranking functions are extended to rate formulas and conditionals in the following way. The *rank* $\kappa(A)$ of a formula A is the minimal rank of its models,

$$\kappa(A) = \min\{\kappa(\omega) \mid \omega \in \Omega(\Sigma),\ \omega \models A\},$$

where the convention $\min \emptyset = \infty$ applies. The rank of a conditional $(B|A)$ is $\kappa(B|A) = \kappa(AB) - \kappa(A)$. A ranking function κ *accepts* a conditional $(B|A)$, written as $\kappa \models (B|A)$, iff $\kappa(AB) < \kappa(A\overline{B})$ or $\kappa(A) = \infty$. Eventually, κ is a *(ranking) model* of a *belief base* Δ iff κ accepts all conditionals in Δ.

A belief base is called *consistent* iff it has at least one model. Consistency of a belief base can be checked based on the notion of *tolerance* [17]. A belief base Δ *tolerates* a conditional $(B|A)$ iff there is a possible world $\omega \in \Omega(\Sigma)$ such that ω *verifies* $(B|A)$, i.e., $\omega \models AB$, and no conditional from Δ is *falsified* by ω, i.e., $\omega \models (A' \Rightarrow B')$ for all $(B'|A') \in \Delta$. With this, the consistency check of Δ goes as follows (cf. [8]): Remove all conditionals from Δ which are tolerated by Δ. Then, repeat with the reduced belief base. If this procedure ends up in the empty belief base $\emptyset$, then Δ is consistent. Otherwise, Δ is inconsistent.

The set of all consistent belief bases over Σ is denoted by $\mathcal{D}(\Sigma)$. As for formulas, we denote the set of atoms which are mentioned in X with $\Sigma(X)$, whether X is a single conditional or a whole belief base.

Example 1. *In Table 1 an example of a belief base about an animal world is shown. The signature of this belief base $\Delta^{\mathfrak{a}}$ is $\Sigma(\Delta^{\mathfrak{a}}) = \{a, b, c, d, f, h, i, k, l, m, p, r, s, w\}$ and the signature of conditional $\mathfrak{r}_6$ is $\Sigma(\mathfrak{r}_6) = \{b, p\}$, for example. The proof that $\Delta^{\mathfrak{a}}$ is consistent is straightforward (please see also Table 3 and Example 5).*

2.3 Inductive Inference

We consider the task of drawing *inductive inferences* from a consistent belief base Δ. Roughly said, we understand inductive inferences as conditionals $\mathfrak{q} = (B|A) \in \mathcal{CL}(\Sigma)$ which are *plausible consequences* from Δ. In other words, if one accepts Δ, then one should presumably accept $\mathfrak{q}$ either. In order to formally define what a plausible consequence from Δ is, we consider *(inductive) inference operators* as in [11].

Conditional	Meaning
$\mathfrak{r}_1 = (f\|aw)$	Winged (w) animals (a) usually fly (f).
$\mathfrak{r}_2 = (\bar{f}\|a\bar{w})$	Wingless animals usually do not fly.
$\mathfrak{r}_3 = (b \Rightarrow a\|\top)$	Birds (b) are animals.
$\mathfrak{r}_4 = (w\|b)$	Birds usually have wings.
$\mathfrak{r}_5 = (d\|b)$	Birds usually drink (d) water.
$\mathfrak{r}_6 = (p \Rightarrow b\|\top)$	Penguins (p) are birds.
$\mathfrak{r}_7 = (\bar{f}\|p)$	Penguins usually do not fly.
$\mathfrak{r}_8 = (c \Rightarrow b\|\top)$	Chicken (c) are birds.
$\mathfrak{r}_9 = (\bar{f}\|c)$	Chicken usually do not fly.
$\mathfrak{r}_{10} = (f\|cs)$	Scared (s) chicken usually fly.
$\mathfrak{r}_{11} = (\bar{s}\|c)$	Chicken are usually not scared.
$\mathfrak{r}_{12} = (i \Rightarrow a\|\top)$	Fish (i) are animals.
$\mathfrak{r}_{13} = (r \Rightarrow i\|\top)$	Freshwater fish (r) are fish.
$\mathfrak{r}_{14} = (l \Rightarrow i\|\top)$	Saltwater fish (l) are fish.
$\mathfrak{r}_{15} = (l \vee r\|i)$	Fish are usually saltwater fish or freshwater fish.
$\mathfrak{r}_{16} = (\bar{d}\|r)$	Freshwater fish usually do not drink water.
$\mathfrak{r}_{17} = (d\|l)$	Saltwater fish usually drink water.
$\mathfrak{r}_{18} = (h \Rightarrow r\|\top)$	Hatchetfish (h) are freshwater fish.
$\mathfrak{r}_{19} = (f\bar{w}\|h)$	Hatchetfish usually fly but are wingless.
$\mathfrak{r}_{20} = (k \Rightarrow m\|\top)$	Kangaroos (k) are marsupials (m).

Table 1: Belief base $\Delta^a = \{\mathfrak{r}_1, \ldots, \mathfrak{r}_{20}\}$ from Example 1.

Definition 1 ((Inductive) Inference Operator). *Let Σ be a signature. An (inductive) inference operator $\mathfrak{I} : \mathcal{D}(\Sigma) \to 2^{\mathcal{L}(\Sigma) \times \mathcal{L}(\Sigma)}$ is a mapping which assigns to each consistent belief base $\Delta \in \mathcal{D}(\Sigma)$ an inference relation $\mathrel{\vert\!\sim}^{\mathfrak{I}}_\Delta \subseteq \mathcal{L}(\Sigma) \times \mathcal{L}(\Sigma)$ such that:*

- *If $(B|A) \in \Delta$, then $A \mathrel{\vert\!\sim}^{\mathfrak{I}}_\Delta B$.* (Direct Inference)

- *If $\Delta = \emptyset$, then $A \mathrel{\vert\!\sim}^{\mathfrak{I}}_\Delta B$ only if $A \models B$.* (Trivial Vacuity)

With a slight abuse of notation, we denote with $\mathfrak{I}(\Delta) = \{(B|A) \in \mathcal{CL}(\Sigma) \mid A \mathrel{\vert\!\sim}^{\mathfrak{I}}_\Delta B\}$ the set of all inductive inferences which can be drawn from Δ with respect to $\mathfrak{I}$.

The term *inductive* is used because inductive inference operators are able to generate new inferences from a set of given conditional beliefs Δ. In our case, they are even capable of completing belief bases to whole belief states, i.e., ranking functions. Note that the inference relations $\mathrel{\vdash\mkern-9mu\sim}_\Delta^{\mathfrak{I}}$ are parametrized by Δ in the sense that they use Δ as a base for an inductive generation process while respecting Direct Inference and Trivial Vacuity. Inductive inference leads to a three-valued *inference response* to a *query conditional* $(B|A) \in \mathcal{CL}(\Sigma)$, going back to de Finetti [6]:

$$[\![(B|A)]\!]_\Delta^{\mathfrak{I}} = \begin{cases} \text{yes} & \text{iff} \quad (B|A) \in \mathfrak{I}(\Delta) \\ \text{no} & \text{iff} \quad (\overline{B}|A) \in \mathfrak{I}(\Delta) \\ \text{unknown} & \text{otherwise} \end{cases}.$$

We now discuss some important representatives of inference operators.

System P

The *System P inference operator* $\mathfrak{I}^P$ [1, 13] is defined by $(B|A) \in \mathfrak{I}^P(\Delta)$ iff *every* model of Δ accepts the conditional $(B|A) \in \mathcal{CL}(\Sigma)$. It is a semantical characterization of a collection of well-established inference rules from nonmonotonic reasoning [1; 13] which are recalled in the next definition.

Definition 2 (System P). *Let $\mathrel{\vdash\mkern-9mu\sim} \subseteq \mathcal{L}(\Sigma) \times \mathcal{L}(\Sigma)$ be an inference relation, and let $A, B, C \in \mathcal{L}(\Sigma)$ be formulas. Then, System P is the collection of the following inference rules:*

$$\begin{aligned}
& A \mathrel{\vdash\mkern-9mu\sim} A, && \text{(Reflexivity)} \\
AB \mathrel{\vdash\mkern-9mu\sim} C \text{ and } A \mathrel{\vdash\mkern-9mu\sim} B \quad \text{imply} \quad & A \mathrel{\vdash\mkern-9mu\sim} C, && \text{(Cut)} \\
A \mathrel{\vdash\mkern-9mu\sim} B \text{ and } A \mathrel{\vdash\mkern-9mu\sim} C \quad \text{imply} \quad & AB \mathrel{\vdash\mkern-9mu\sim} C, && \text{(Cautious Monotony)} \\
A \mathrel{\vdash\mkern-9mu\sim} B \text{ and } B \models C \quad \text{imply} \quad & A \mathrel{\vdash\mkern-9mu\sim} C, && \text{(Right Weakening)} \\
A \mathrel{\vdash\mkern-9mu\sim} C \text{ and } B \mathrel{\vdash\mkern-9mu\sim} C \quad \text{imply} \quad & A \vee B \mathrel{\vdash\mkern-9mu\sim} C, && \text{(Or)} \\
A \equiv B \text{ and } B \mathrel{\vdash\mkern-9mu\sim} C \quad \text{imply} \quad & A \mathrel{\vdash\mkern-9mu\sim} C. && \text{(Left Logical Equivalence)}
\end{aligned}$$

Whether $(B|A) \in \mathfrak{I}^P(\Delta)$ holds or not can be decided based on a consistency check. One has $(B|A) \in \mathfrak{I}^P(\Delta)$ iff $\Delta \cup \{(\overline{B}|A)\}$ is inconsistent [8].

Example 2. *From $\Delta^{\mathfrak{a}}$ (cf. Table 1) we can infer that winged birds usually drink water, i.e., $(d|bw) \in \mathfrak{I}^P(\Delta^{\mathfrak{a}})$, because $\Delta^{\mathfrak{a}} \cup \{(\overline{d}|bw)\}$ is inconsistent. An alternative way to prove this inference is to apply the inference rules from System P directly. Since $\mathrel{\vdash\mkern-9mu\sim}_{\Delta^{\mathfrak{a}}}^{\mathfrak{I}^P}$ is an inductive inference operator, it satisfies Direct Inference and we have $b \mathrel{\vdash\mkern-9mu\sim}_{\Delta^{\mathfrak{a}}}^{\mathfrak{I}^P} w$ as well as $b \mathrel{\vdash\mkern-9mu\sim}_{\Delta^{\mathfrak{a}}}^{\mathfrak{I}^P} d$. Consequently, with Cautious Monotony we can infer $bw \mathrel{\vdash\mkern-9mu\sim}_{\Delta^{\mathfrak{a}}}^{\mathfrak{I}^P} d$.*

A meaningful property of inductive inference operators in the context of activation-based conditional inference is *Semi-Monotony*. An inductive inference operator $\mathfrak{I}$ is called *semi-monotonous* iff for every two consistent belief bases Δ and $\tilde{\Delta}$ it holds that $\tilde{\Delta} \subseteq \Delta$ implies $\mathfrak{I}(\tilde{\Delta}) \subseteq \mathfrak{I}(\Delta)$. Note that for Semi-Monotony, the idea of induction is crucial because Semi-Monotony means monotony with respect to the parameter Δ, while normal monotony focuses on the premises A of inferences $A \hspace{0.1em}\sim\hspace{-0.9em}\mid\hspace{0.4em} B$. A proof that the System P inference operator $\mathfrak{I}^P$ is *semi-monotonous* is given in [23], but this is also obvious from the syntactic characterization via the inference rules in Definition 2. Here, we give an example which illustrates the Semi-Monotony of System P.

Example 3. *Consider $\Delta^{\mathfrak{a}'} = \{\mathfrak{r}_9, \mathfrak{r}_{11}\} \subseteq \Delta^{\mathfrak{a}}$ (cf. Table 1). Since $\Delta^{\mathfrak{a}'} \cup \{(f|c\bar{s})\}$ is inconsistent, $c\bar{s} \hspace{0.1em}\sim\hspace{-0.9em}\mid\hspace{0.4em}^P_{\Delta^{\mathfrak{a}'}} \bar{f}$ follows. That is, one can infer from $\Delta^{\mathfrak{a}'}$ with respect to System P that chicken which are not scared usually do not fly. Due to the Semi-Monotony of System P, this inference can also be drawn from $\Delta^{\mathfrak{a}}$ because of $\Delta^{\mathfrak{a}'} \subseteq \Delta^{\mathfrak{a}}$.*

System Z

Another well-known inference operator is provided by *System Z* [17]. The *System Z inference operator* $\mathfrak{I}^Z$ makes use of the so-called *Z-partition* of a consistent belief base Δ as an auxiliary structure for computing $\mathfrak{I}^Z(\Delta)$. Z-partitions are specific ordered partitions of belief bases. An ordered partition $(\Delta_0, \Delta_1, \ldots, \Delta_m)$ of Δ is called *tolerance partition* of Δ iff, for $i = 0, \ldots, m$, every conditional in Δ_i is tolerated by $\bigcup_{j=i}^m \Delta_j$. The *Z-partition* $Z(\Delta)$ is the unique tolerance partition of Δ which is obtained by iteratively determining Δ_i as the ˌmaximal set of tolerated conditionals. If a conditional $\mathfrak{r}$ is in the i-th partition of $Z(\Delta)$, we say that $\mathfrak{r}$ has *Z-rank* $Z^\Delta(\mathfrak{r}) = i$. Therewith, a conditional $(B|A) \in \mathcal{CL}(\Sigma)$ is inferred from a consistent belief base Δ with respect to System Z, written $(B|A) \in \mathfrak{I}^Z(\Delta)$, iff the ranking model κ^Z_Δ of Δ which is defined by

$$\kappa^Z_\Delta(\omega) = \begin{cases} 0, & \text{iff } \forall (B'|A') \in \Delta : \omega \models A' \Rightarrow B' \\ \max\{Z^\Delta(\mathfrak{r}) \mid \mathfrak{r} = (B'|A') \in \Delta : \omega \models A'\overline{B'}\} + 1, & \text{otherwise} \end{cases}$$

accepts $(B|A)$.

Drawing inferences in System Z corresponds precisely to the entailment by *rational closure* [15, 7]. While $\mathfrak{I}^Z$ is not semi-monotonous, it satisfies the property *Rational Monotonicity* in contrast to the System P inference operator $\mathfrak{I}^P$. Let $A, B, C \in \mathcal{L}(\Sigma)$, then an inference relation $\hspace{0.1em}\sim\hspace{-0.9em}\mid\hspace{0.4em} \subseteq \mathcal{L}(\Sigma) \times \mathcal{L}(\Sigma)$ satisfies *Rational*

Cond.	$Z^{\Delta^a}(r_i)$	Cond.	$Z^{\Delta^a}(r_i)$	Cond.	$Z^{\Delta^a}(r_i)$	Cond.	$Z^{\Delta^a}(r_i)$
r_1	0	r_6	0	r_{11}	1	r_{16}	0
r_2	0	r_7	1	r_{12}	0	r_{17}	0
r_3	0	r_8	0	r_{13}	0	r_{18}	0
r_4	0	r_9	1	r_{14}	0	r_{19}	1
r_5	0	r_{10}	2	r_{15}	0	r_{20}	0

Table 2: Z-ranks $Z^{\Delta^a}(r)$ of the conditionals in the belief base Δ^a (cf. Table 1).

Monotonicity iff

$$A \mathrel{\vert\!\sim} B \text{ and not } A \mathrel{\vert\!\sim} \overline{C} \quad \text{imply} \quad AC \mathrel{\vert\!\sim} B. \qquad \text{(Rational Monotonicity)}$$

In the following example, we apply System Z inference to the belief base Δ^a from Example 1. All these System Z inferences as well as the ordered partition of Δ^a employed by System Z can easily be computed using the online system InfOCF-Web [14] which also provides implementations of inferences according to System P as well as implementations of other nonmonotonic inference operators.

Example 4. *We again consider the belief base Δ^a from Table 1. The Z-ranks of the conditionals in Δ^a are shown in Table 2. For example, the Z-rank of r_1 is $Z^{\Delta^a}(r_1) = 0$ because r_1 is tolerated by Δ^a (consider $\omega = a\bar{b}\bar{c}df\bar{h}\bar{i}\bar{k}\bar{l}\bar{m}\bar{p}\bar{r}\bar{s}w$). According to System Z, we have $p \mathrel{\vert\!\sim}^{\mathfrak{J}^Z}_{\Delta^a} \bar{f}$ because $(\bar{f}|p) \in \Delta^a$ and, as being an inductive inference operator, $\mathfrak{J}^Z$ satisfies Direct Inference. Together with the fact that it cannot be inferred from Δ^a that penguins usually do not have wings, i.e., $p \mathrel{\not\vert\!\sim}^{\mathfrak{J}^Z}_{\Delta^a} \bar{w}$, Rational Monotonicity tells us that winged penguins usually do not fly, in symbols $pw \mathrel{\vert\!\sim}^{\mathfrak{J}^Z}_{\Delta^a} \bar{f}$ resp. $(\bar{f}|pw) \in \mathfrak{J}^Z(\Delta^a)$. Note that this inference cannot be drawn in System P. The fact that $p \mathrel{\not\vert\!\sim}^{\mathfrak{J}^Z}_{\Delta^a} \bar{w}$ holds can be shown as follows: Because $\omega = a\bar{b}\bar{c}df\bar{h}\bar{i}\bar{k}\bar{l}\bar{m}\bar{p}\bar{r}\bar{s}w$ falsifies only conditional r_7 and the Z-rank of r_7 is $Z^{\Delta^a}(r_7) = 0$, one has, on the one hand, $\kappa^Z_{\Delta^a}(pw) \leq 1$. On the other hand, one has $\kappa^Z_{\Delta^a}(p\bar{w}) \geq 1$ because every possible world ω which models $p\bar{w}$ either falsifies at least r_4 (if $\omega \models b$) or at least r_6 (if $\omega \models \bar{b}$). Together, $\kappa^Z_{\Delta^a}(pw) \leq \kappa^Z_{\Delta^a}(p\bar{w})$ follows, which implies $p \mathrel{\not\vert\!\sim}^{\mathfrak{J}^Z}_{\Delta^a} \bar{w}$.*

A further property of System Z inference which will be of importance here is that the Z-partition $Z(\Delta)$ of the conditionals in Δ satisfies the paradigm of *maximum normality*, i.e., the lower the Z-rank of a conditional is, the more normal the conditional is.

Example 5. *Once again we refer to the belief base Δ^a from Table 1 and the Z-ranks of the conditionals in Δ^a according to Table 2. The Z-ranks $Z^{\Delta^a}(r_{10}) = 2$*

and $Z^{\Delta^a}(\mathfrak{r}_9) = 1$ *illustrate the concept of normality well. While conditional* $\mathfrak{r}_9$ *is concerned about the flight ability of chicken in general, conditional* $\mathfrak{r}_{10}$ *makes a statement about the flight behavior of chicken when they are in a special mood. Hence, conditional* $\mathfrak{r}_{10}$ *applies to a more specific case than* $\mathfrak{r}_9$ *and accordingly has a higher Z-rank.*

c-Representations

We finally want to bring the concept of *c-representations* [10] into focus. c-Representations are special ranking functions obtained by assigning individual integer impacts to the conditionals in Δ. In a second step, the ranks of the possible worlds are generated as the sum of impacts of falsified conditionals.

Definition 3 (c-Representation). *A c-representation* κ_Δ^c *of a consistent belief base* $\Delta = (B_1|A_1), \ldots, (B_n|A_n)$ *is a ranking function constructed from non-negative integer impacts* $\eta_j \in \mathbb{N}_0$ *assigned to each* $(B_j|A_j) \in \Delta$ *such that* κ_Δ^c *accepts* Δ *and is given by*

$$\kappa_\Delta^c(\omega) = \sum_{\substack{j=1,\ldots,n: \\ \omega \models A_j \overline{B_j}}} \eta_j.$$

The c-representations of a consistent belief base Δ constitute a family of ranking functions, depending on the concrete specification of the impact values η_j. Any c-representation κ_Δ^c of this family allows one to draw inductive inferences from Δ. In order to obtain an inference operator, one has to fix a single c-representation per belief base, preferably based on a common methodology. One possible way of doing so is to consider the so-called Z-c-representations [10] which are c-representations that make use of the System-Z partition to compute the impact values η_j in a unique way.

Here, we mention c-representations because drawing inferences based on c-representation (c-inference) satisfies *Syntax Splitting* [11]. The idea behind syntax splitting is that if a belief base splits into two syntactically independent parts, then reasoning about one part of the belief base should be independent of the other part. Like Semi-Monotony, Syntax Splitting is obviously a property of interest when identifying subsets of a consistent belief base Δ from which meaningful inferences or even the same inferences as from Δ can be drawn. For the technical details, see [11].

2.4 Focused Inference

Taking up the idea of identifying meaningful subsets of a belief base for drawing inferences, we now recall the concept of *focused inference* from [23]. The idea behind focused inference is to draw inferences from a reasonable (as small as possible) subset of Δ in order to make snap but still well-founded decisions in time. In this context, the advantage of semi-monotonous inference operators like $\mathfrak{I}^P$ is that one does not risk to draw false inferences when focusing on a subset $\tilde{\Delta} \subseteq \Delta$ because $[\![(B|A)]\!]_{\tilde{\Delta}}^{\mathfrak{I}^P} = $ yes (resp. no) implies $[\![(B|A)]\!]_{\Delta}^{\mathfrak{I}^P} = $ yes (resp. no). In order to formalize focused inference, we consider mappings $\phi : \mathcal{D}(\Sigma) \to \mathcal{D}(\Sigma)$ with $\phi(\Delta) \subseteq \Delta$, i.e. mappings which return subsets of Δ. We call such a mapping ϕ a *focus*.

Definition 4 (Focused Inference)**.** *Let Δ be a belief base, $(B|A)$ a conditional, $\mathfrak{I}$ an inference operator, and ϕ a focus. Then, $(B|A)$ follows from Δ wrt. $\mathfrak{I}$ in the focus ϕ iff $(B|A) \in \mathfrak{I}(\phi(\Delta))$.*

In [23], the focus ϕ is defined iteratively based on the query $\mathfrak{q} = (B|A)$: The conditionals in the *direct focus* $\phi_0^{\mathfrak{q}}$ are those conditionals which share at least one atom with $\mathfrak{q}$, i.e. $\phi_0^{\mathfrak{q}}(\Delta) = \{\mathfrak{r} \in \Delta \mid \Sigma(\mathfrak{r}) \cap \Sigma(\mathfrak{q}) \neq \emptyset\}$. The conditionals in the *i*-th focus are determined by $\phi_i^{\mathfrak{q}}(\Delta) = \{\mathfrak{r} \in \Delta \mid \exists \mathfrak{r}' \in \phi_{i-1}^{\mathfrak{q}}(\Delta) : \Sigma(\mathfrak{r}) \cap \Sigma(\mathfrak{r}') \neq \emptyset\}$. Note that this definition of the foci respects the concept of Syntax Splitting. That is, if the belief base Δ splits into syntactically independent subsets Δ_1 and Δ_2 with $\Delta_i \subseteq \mathcal{CL}(\Sigma_i)$ for $i = 1, 2$ and $\Sigma_1 \cap \Sigma_2 = \emptyset$, and if $\mathfrak{q} \in \mathcal{CL}(\Sigma_1)$, then $\phi_i^{\mathfrak{q}}(\Delta) \subseteq \Delta_1$ for all $i \in \mathbb{N}_0$. Analogously, $\phi_i^{\mathfrak{q}}(\Delta) \subseteq \Delta_2$ if $\mathfrak{q} \in \mathcal{CL}(\Sigma_2)$.

Example 6. *The belief base $\Delta^{\mathfrak{a}}$ (cf. Table 1) splits into the syntactically independent subsets $\Delta_1^{\mathfrak{a}} = \{\mathfrak{r}_1, \ldots, \mathfrak{r}_{19}\}$ and $\Delta_2^{\mathfrak{a}} = \{\mathfrak{r}_{20}\}$. The direct focus of $\Delta^{\mathfrak{a}}$ wrt. $\mathfrak{q} = (f|c\bar{s})$ is $\Delta_0 = \phi_0^{\mathfrak{q}}(\Delta^{\mathfrak{a}})$ with $\Delta_0 = \{\mathfrak{r}_1, \mathfrak{r}_2, \mathfrak{r}_7, \mathfrak{r}_8, \mathfrak{r}_9, \mathfrak{r}_{10}, \mathfrak{r}_{11}, \mathfrak{r}_{19}\}$. One can show that $[\![\mathfrak{q}]\!]_{\Delta_0}^{\mathfrak{I}^P} = $ no. According to Example 3, one already has $[\![\mathfrak{q}]\!]_{\Delta^{\mathfrak{a}'}}^{\mathfrak{I}^P} = $ no for $\Delta^{\mathfrak{a}'} = \{\mathfrak{r}_9, \mathfrak{r}_{11}\} \subset \Delta_0$, though. Hence, the direct focus according to [23] does not have to be the smallest possible focus in which an inference can be drawn. On the contrary, a focus can also be too small in order to decide a query. For instance, $[\![\mathfrak{q}]\!]_{\Delta^{\mathfrak{a}''}}^{\mathfrak{I}^P} = $ unknown with respect to any $\Delta^{\mathfrak{a}''} \subset \Delta^{\mathfrak{a}}$ with $\{\mathfrak{r}_9, \mathfrak{r}_{11}\} \not\subseteq \Delta^{\mathfrak{a}''}$.*

To sum up, apart from the computational benefits of drawing inferences with respect to small foci, appropriate foci are also interesting from the knowledge representation and reasoning (KRR) perspective because they may unveil the part of the belief base which is relevant for answering the query.

3 Cognitive Foundations

We now approach activation-based conditional inference from the cognitive perspective and discuss the *ACT-R architecture*. In particular, we focus on the activation process of chunks.

3.1 ACT-R Architecture

ACT-R (Adaptive Control of Thought-Rational, [5, 4]) is a production systems based cognitive architecture with the aim to formalize human reasoning. In ACT-R a distinction is made between the *declarative* and the *procedural* memory. In the declarative memory, categorical knowledge about individuals or objects is stored in form of *chunks* (*knowing that*) while the procedural memory consists of *production rules* and describes how the chunks are processed (*knowing how*, [19]).

Reasoning in ACT-R starts with an *initial priming*, for example a stimulus from the environment, which causes an activation of chunks. The chunk with the highest activation is processed by a selection of production rules in order to compute a solution to the reasoning task. If this fails, the activation passes into an iterative process: The system obtains additional chunks from the declarative memory and tries to compute a solution again. The iteration stops when either the problem is solved or no further chunks are active.

The retrieval of chunks is a very refined process in ACT-R. Basically, it depends on an *activation function* which is calculated for each specific request anew and which is based on a *usage history* of the chunks, associations between *cognitive units* and the *priming* [3]. There is no clear consensus about the kind of cognitive units despite of the perception that they form the basic building blocks of thinking [2].

3.2 Activation of Chunks

How the activation of a chunk $\mathcal{A}(\mathfrak{c}_i)$ is computed in detail depends on multiple parameters and the configurations of the ACT-R system, but is mainly given by the sum of the so-called *base-level activation* $\mathcal{B}(\mathfrak{c}_i)$ and the *spreading activation* $\mathcal{S}(\mathfrak{c}_i)$, which again is a sum of *degrees of association* between chunks $\mathcal{S}(\mathfrak{c}_i, \mathfrak{c}_j)$ weighted by some *weighting factors* $\mathcal{W}(\mathfrak{c}_j)$:

$$\mathcal{A}(\mathfrak{c}_i) = \underbrace{\mathcal{B}(\mathfrak{c}_i)}_{\text{base-level activation}} + \underbrace{\sum_j \mathcal{W}(\mathfrak{c}_j) \cdot \mathcal{S}(\mathfrak{c}_i, \mathfrak{c}_j)}_{\text{spreading activation } \mathcal{S}(\mathfrak{c}_i)} \ . \qquad (1)$$

The *base-level activation* of a chunk $\mathcal{B}(\mathfrak{c}_i)$ reflects the *entrenchment* of $\mathfrak{c}_i$ in the reasoner's memory and depends on the recency and frequency of its use. Typically, $\mathcal{B}(\mathfrak{c}_i)$ is decreased over time (*fading out*) and is increased when the chunk is active. Further, $\mathcal{B}(\mathfrak{c}_i)$ is independent of the priming.

In contrast, the *spreading activation* of a chunk $\mathcal{S}(\mathfrak{c}_i)$ depends on the priming and exploits the well-known *spreading activation theory* [3] to formalize how the brain iterates through a network of associated ideas to retrieve information. In the spreading activation theory one breaks down the notion of ideas into *cognitive units*. Usually, the cognitive units are arranged as vertices in an undirected graph, the so-called *spreading activation network* $\mathcal{N}(\Delta)$, and an initial *triggering* of some cognitive units caused by the priming is propagated through $\mathcal{N}(\Delta)$. The spreading activation $\mathcal{S}(\mathfrak{c}_i)$ can then be derived from the *triggering values* of the cognitive units of which $\mathfrak{c}_i$ makes use. The interrelation of cognitive units and of chunks is specified in more detail in the *degree of association* and the *weighting factor*.

The *degree of association* $\mathcal{S}(\mathfrak{c}_i, \mathfrak{c}_j)$ reflects how strongly related $\mathfrak{c}_i$ and $\mathfrak{c}_j$ are. Chunks which deal with the same issue have a high degree of association while chunks which refer to different topics are only loosely or not related and, therefore, have a low degree of association. Technically, $\mathcal{S}(\mathfrak{c}_i, \mathfrak{c}_j)$ is based on the cognitive units which $\mathfrak{c}_i$ and $\mathfrak{c}_j$ have in common. The degrees of association are weighted by the *weighting factors* $\mathcal{W}(\mathfrak{c}_i)$. While the degree of association is independent of the priming, the weighting factors reflect the context-dependency of $\mathcal{A}(\mathfrak{c}_i)$. Only if $\mathfrak{c}_i$ is associated to a chunk $\mathfrak{c}_j$ ($\mathcal{S}(\mathfrak{c}_i, \mathfrak{c}_j) > 0$) which has positive weight ($\mathcal{W}(\mathfrak{c}_j) > 0$), then the chunk $\mathfrak{c}_i$ has a positive spreading activation ($\mathcal{S}(\mathfrak{c}_i) > 0$), too.

4 Activation-Based Conditional Inference

We now apply the ACT-R machinery to conditional reasoning and develop *activation-based conditional inference*. We start with a general integration of the activation of conditionals into the focused inference pipeline, before we discuss the activation function for conditionals in detail. We conclude this section by showing how cognitive concepts such as *forgetting* and *remembering* fit into our activation-based conditional inference approach.

4.1 Integration of the Activation of Conditionals into Focused Inference

As common ACT-R implementations are production systems which process chunks that are represented as simple lists of attributes, the logical basis of ACT-R does not hold the pace with modern KRR formalisms in nonmonotonic reasoning. Thus, we propose a cognitively inspired model of inductive conditional reasoning by interpreting the concepts of ACT-R in terms of logic, conditionals, and inference. More precisely, we replace chunks by the conditionals of a belief base Δ and derive a focus ϕ based on the activation function in Equation (1) in order to draw focused inferences with respect to any inference operator $\mathfrak{I}(\phi(\Delta))$. Here, we rely on $\mathfrak{I}^P(\phi(\Delta))$ because of the semi-monotony of System P. In our formalism, atoms play the role of cognitive units, and the production rules are replaced by the inference operator.

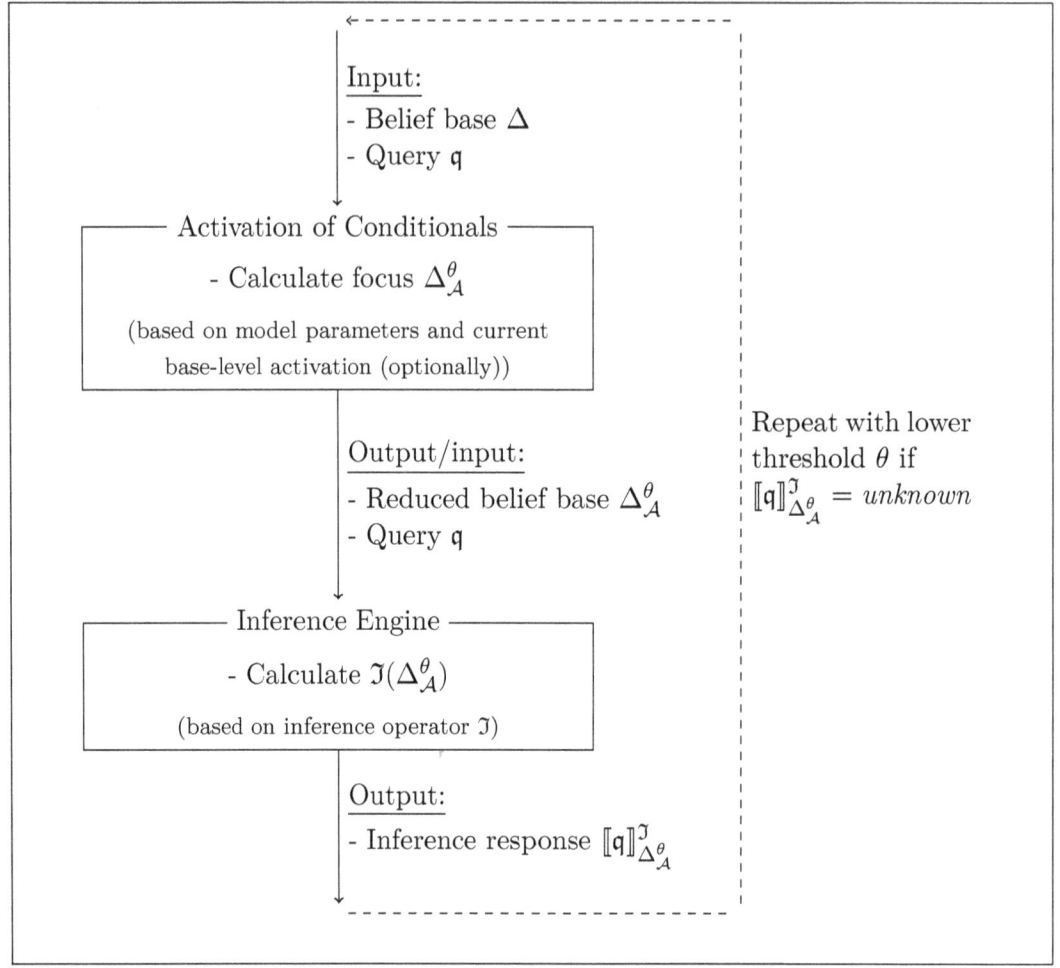

Figure 1: Activation-based conditional inference pipeline.

From the conditional logical perspective, the added value of this *activation-based conditional inference* approach are

- the cognitive justification of the focus,
- the possibility of a more fine-grained adjustment of the focus than in [23],
- and the option to integrate further cognitive concepts such as forgetting and remembering.

Formally, we calculate an activation value $\mathcal{A}(\mathfrak{r}) > 0$ for every conditional $\mathfrak{r}$ in Δ. If this activation $\mathcal{A}(\mathfrak{r})$ is above a certain threshold, say $\mathcal{A}(\mathfrak{r}) \geq \theta$, the conditional is

selected for the focus $\phi(\Delta)$. For this, we consider a *selection function* $s_{\mathcal{A}}^{\theta} : \Delta \to \{0,1\}$ with $s_{\mathcal{A}}^{\theta}(\mathfrak{r}) = 1$ iff $\mathcal{A}(\mathfrak{r}) \geq \theta$ (and $s_{\mathcal{A}}^{\theta}(\mathfrak{r}) = 0$ otherwise) and denote the set of selected conditionals by
$$\Delta_{\mathcal{A}}^{\theta} = \{\mathfrak{r} \in \Delta \mid s_{\mathcal{A}}^{\theta}(\mathfrak{r}) = 1\}.$$
Possible strategies for choosing suitable thresholds θ are (a) guaranteeing that a certain percentage of conditionals is selected or (b) maximizing the gap between the activation value of the least activated selected conditional and the most activated unselected conditional. Note that $\Delta_{\mathcal{A}}^{\theta}$ will implicitly depend on a query conditional $\mathfrak{q} = (B|A)$ since queries will serve as the initial priming and the spreading activation, which is part of $\mathcal{A}$, depends on this priming.

Definition 5 (Activation-Based Conditional Inference). *Let Δ be a belief base, $(B|A)$ a conditional, $\mathfrak{I}$ an inference operator, $\mathcal{A}$ an activation function for Δ, and let $\theta \geq 0$. Then, $(B|A)$ is activation-based inferred from Δ wrt. $\mathfrak{I}$, $\mathcal{A}$, and θ iff $(B|A) \in \mathfrak{I}(\Delta_{\mathcal{A}}^{\theta})$.*

If answering a query fails, i.e. $[\![(B|A)]\!]_{\Delta_{\mathcal{A}}^{\theta}}^{\mathfrak{I}} = \textit{unknown}$, then the inference process can be repeated by iteratively choosing a lower threshold $\theta_{i+1} < \theta_i$ which leads to a larger (or equal) set of selected conditionals. In the limit, when choosing $\theta = 0$, one has $\Delta_{\mathcal{A}}^{0} = \Delta$, thus $\mathfrak{I}_{\Delta_{\mathcal{A}}^{0}} = \mathfrak{I}_{\Delta}$. This iteration process is in analogy to the sequence $(\phi_i^{\mathfrak{q}}(\Delta))_{i \in \mathbb{N}_0}$ defined in [23] and can be used to approximate $\mathfrak{I}(\Delta)$ for any inductive inference operator $\mathfrak{I}$. In particular, the chance of successfully answering the query increases with each iteration step when $\mathfrak{I}$ is semi-monotonous. The whole inference pipeline is shown in Figure 1.

4.2 Blueprint for Activation-Based Conditional Inference

ACT-R does not formalize the activation function in Equation (1) in detail but describes its functionality informally. Hence, there is certain freedom in its configuration. We give a concrete instantiation of the activation function within the conditional inference setting which can be seen as a blueprint for further investigations and empirical analyses. Note that we shift the dependence of the base-level activation on the usage history of conditionals to the next section.

Let Δ be a belief base, $\mathfrak{r}_i \in \Delta$, and $\mathfrak{q}$ a conditional (the query conditional which serves as the initial priming in this context as well). Then, Equation (1) becomes

$$\mathcal{A}_{\mathfrak{q}}^{\Delta}(\mathfrak{r}_i) = \underbrace{\mathcal{B}^{\Delta}(\mathfrak{r}_i)}_{\text{base-level activation}} + \underbrace{\sum_{\mathfrak{r}_j \in \Delta} \mathcal{W}_{\mathfrak{q}}^{\Delta}(\mathfrak{r}_j) \cdot \mathcal{S}(\mathfrak{r}_i, \mathfrak{r}_j)}_{\text{spreading activation } \mathcal{S}_{\mathfrak{q}}^{\Delta}(\mathfrak{r}_i)}.$$

We explain the single components of $\mathcal{A}_{\mathfrak{q}}^{\Delta}(\mathfrak{r}_i)$ in detail.

Base-Level Activation

The base-level activation $\mathcal{B}^{\Delta}(\mathfrak{r})$ reflects the entrenchment of $\mathfrak{r}$ in the reasoner's memory. Since epistemic entrenchment and ranking semantics are dual ratings, the *normality* of a conditional is a good estimator and we define

$$\mathcal{B}^{\Delta}(\mathfrak{r}) = \frac{1}{1 + Z^{\Delta}(\mathfrak{r})}, \qquad \mathfrak{r} \in \Delta,$$

where $Z^{\Delta}(\mathfrak{r})$ is the Z-rank of $\mathfrak{r}$. Following this definition, $\mathcal{B}^{\Delta}(\mathfrak{r})$ is positive and normalized by 1. While the most normal conditionals have a base-level activation of $\mathcal{B}^{\Delta}(\mathfrak{r}) = 1$, this value decreases with increasing specificity of $\mathfrak{r}$.

Example 7. *Table 3 shows the base-level activations of the conditionals in $\Delta^{\mathfrak{a}}$ (Table 1). For example, $\mathcal{B}^{\Delta^{\mathfrak{a}}}(\mathfrak{r}_9) = 1/2$ and $\mathcal{B}^{\Delta^{\mathfrak{a}}}(\mathfrak{r}_{10}) = 1/3$. Since $\mathfrak{r}_9$ is less specific than $\mathfrak{r}_{10}$ (cf. Example 5), its base-level activation is higher than $\mathcal{B}^{\Delta^{\mathfrak{a}}}(\mathfrak{r}_{10})$.*

When taking no account of the spreading activation $\mathcal{S}_{\mathfrak{q}}^{\Delta}(\mathfrak{r})$ and considering the base-level activation $\mathcal{B}^{\Delta}(\mathfrak{r})$ only, then the selection function $s_{\mathcal{A}}^{\theta}$ selects the conditionals from the first j sets of the System Z-partition $Z(\Delta) = (\Delta_0, \Delta_1, \ldots, \Delta_m)$, where j depends on the threshold θ. For example, when $\theta = 1$, the conditionals from Δ_0 are selected, i.e. $\Delta_{\mathcal{A}}^1 = \Delta_0$, and when $\theta = 0.5$, then $\Delta_{\mathcal{A}}^{0,5} = \Delta_0 \cup \Delta_1$. Note that this closeness to System Z does not necessarily mean that activation-based conditional inferences are similar to System Z inferences. The base-level activation only affects which conditionals are selected for reasoning, the actual inferences also depend on the inference operator that is applied thereafter.

Weighting Factor

The weighting factor $\mathcal{W}_{\mathfrak{q}}^{\Delta}(\mathfrak{r})$ indicates how much the initial priming $\mathfrak{q}$ triggers the conditional $\mathfrak{r}$. We formalize the influence of the priming according to the spreading activation theory by a labeling of the so-called *spreading activation network* $\mathcal{N}(\Delta)$ between cognitive units. In our context, the cognitive units are simply the atoms in Σ and the outcome of $\mathcal{N}(\Delta)$ is a *triggering value* $\tau_{\mathfrak{q}}^{\Delta}(\mathfrak{a}) \in [0, 1]$ which indicates how much $\mathfrak{a}$ is triggered by $\mathfrak{q}$.

We now discuss how the labeling of the spreading activation network works in detail. The spreading activation network $\mathcal{N}(\Delta) = (\mathcal{V}, \mathcal{E})$ is an undirected graph with vertices $\mathcal{V} = \Sigma$. Edges in $\mathcal{E}$ represent associations between the atoms in Σ

Cond.	$Z^{\Delta^a}(\mathfrak{r}_i)$	$\mathcal{B}^{\Delta^a}(\mathfrak{r}_i)$	$\mathcal{W}^{\Delta^a}_{\mathfrak{q}_1}(\mathfrak{r}_i)$	$\mathcal{S}^{\Delta^a}_{\mathfrak{q}_1}(\mathfrak{r}_i)$	$\mathcal{A}^{\Delta^a}_{\mathfrak{q}_1}(\mathfrak{r}_i)$	$\mathcal{W}^{\Delta^a}_{\mathfrak{q}_2}(\mathfrak{r}_i)$	$\mathcal{S}^{\Delta^a}_{\mathfrak{q}_2}(\mathfrak{r}_i)$	$\mathcal{A}^{\Delta^a}_{\mathfrak{q}_2}(\mathfrak{r}_i)$
$\mathfrak{r}_1$	0	1	1/3	1.36	⟦2.36⟧	1/4	1.27	2.27
$\mathfrak{r}_2$	0	1	1/3	1.36	⟦2.36⟧	1/4	1.27	2.27
$\mathfrak{r}_3$	0	1	2/3	1.41	⟦2.41⟧	1/4	0.66	1.66
$\mathfrak{r}_4$	0	1	1/3	1.13	2.13	1/4	0.70	1.70
$\mathfrak{r}_5$	0	1	2/15	0.82	1.82	1/21	0.41	1.41
$\mathfrak{r}_6$	0	1	2/3	1.36	⟦2.36⟧	1/4	0.60	1.60
$\mathfrak{r}_7$	1	1/2	2/3	1.23	1.73	1/4	1.10	1.60
$\mathfrak{r}_8$	0	1	4/15	1.03	2.03	1/4	1.43	⟦2.43⟧
$\mathfrak{r}_9$	1	1/2	4/15	0.93	1.43	1	2.35	⟦2.85⟧
$\mathfrak{r}_{10}$	2	1/3	2/15	0.81	1.14	1	2.61	⟦2.94⟧
$\mathfrak{r}_{11}$	1	1/2	2/15	0.40	0.90	1	2.08	⟦2.58⟧
$\mathfrak{r}_{12}$	0	1	1/3	0.78	1.78	1/21	0.29	1.29
$\mathfrak{r}_{13}$	0	1	1/15	0.29	1.29	1/21	0.12	1.12
$\mathfrak{r}_{14}$	0	1	1/15	0.27	1.27	4/151	0.08	1.08
$\mathfrak{r}_{15}$	0	1	1/15	0.29	1.29	4/151	0.12	1.12
$\mathfrak{r}_{16}$	0	1	1/15	0.19	1.19	1/21	0.11	1.11
$\mathfrak{r}_{17}$	0	1	1/15	0.17	1.17	4/151	0.07	1.07
$\mathfrak{r}_{18}$	0	1	1/15	0.18	1.18	1/21	0.15	1.15
$\mathfrak{r}_{19}$	1	1/2	1/5	0.89	1.39	1/4	1.09	1.59
$\mathfrak{r}_{20}$	0	1	0	0	1	0	0	1

Table 3: Z-ranks Z^{Δ^a}, base-level activation $\mathcal{B}^{\Delta^a}$, weighting factors $\mathcal{W}^{\Delta^a}_{\mathfrak{q}_i}$, spreading activation $\mathcal{S}^{\Delta^a}_{\mathfrak{q}_i}$, and activation function $\mathcal{A}^{\Delta^a}_{\mathfrak{q}_i}$ wrt. $\mathfrak{q}_1 = (p \Rightarrow a | \top)$ and $\mathfrak{q}_2 = (\bar{f} | c\bar{s})$ for the conditionals in Δ^a. For selected conditionals, the respective degrees of activation are boxed (threshold $\theta = 2.3$).

along which the triggering of the atoms spreads. Two atoms are associated if they occur commonly in some conditionals in Δ, i.e.

$$\mathcal{E} = \{\{a, b\} \mid \exists \mathfrak{r} \in \Delta : \{a, b\} \subseteq \Sigma(\mathfrak{r})\}.$$

The actual spreading of activation is modeled by iteratively labeling the vertices (atoms) in $\mathcal{N}(\Delta)$ with their triggering value $\tau^\Delta_\mathfrak{q}(a)$. The *labeling algorithm* is shown in Figure 3. It starts with labeling the atoms which are mentioned in the query $\mathfrak{q}$ with 1. In the subsequent step, the neighboring atoms are labeled and so on. The remaining atoms which are not reachable from the initially labeled atoms in $\Sigma(\mathfrak{q})$

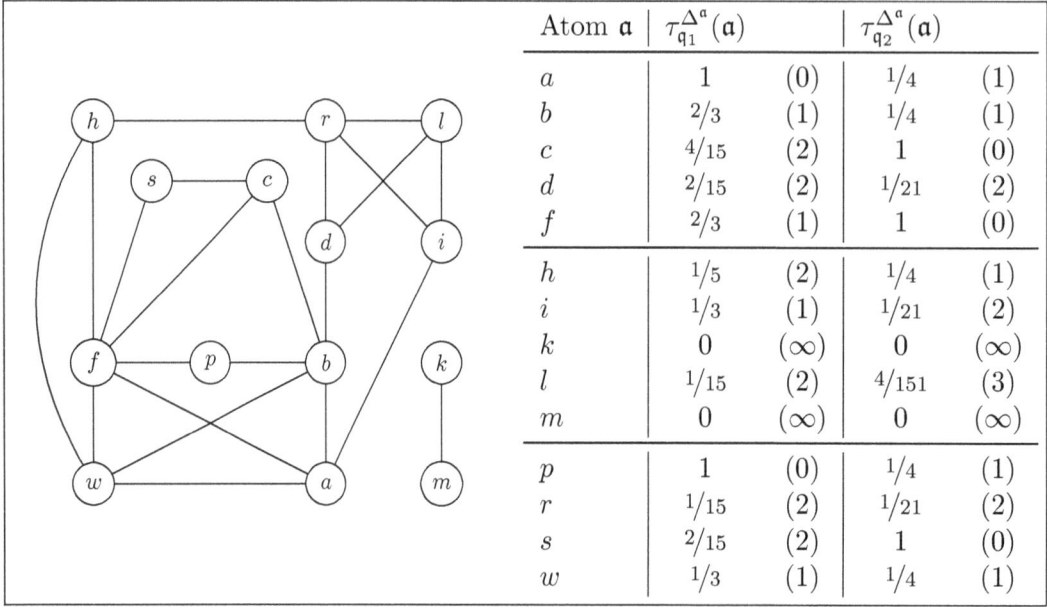

Figure 2: Unlabeled spreading activation network $\mathcal{N}(\Delta^\mathfrak{a})$ and labeling of $\mathcal{N}(\Delta^\mathfrak{a})$ with respect to the queries $\mathfrak{q}_1 = (p \Rightarrow a | \top)$ and $\mathfrak{q}_2 = (f | c\bar{s})$. The numbers in the parentheses next to the labels (i.e., triggering values) are the iteration steps in which the atoms are labeled. 0 stands for the priming and ∞ for unreachable atoms.

are labeled with 0. The labels of the atoms in between are the sum of the labels of the already labeled neighbors weighted by the sum of all labels so far plus 1. This guarantees that these labels are between 0 and 1 and decrease for increasing iteration steps. Therewith, the triggering value of an atom depends on both the count and the triggering values of the associated atoms, i.e. the atoms which are already triggered and which are neighbored in $\mathcal{N}(\Delta)$. The more triggered atoms an atom $\mathfrak{a}$ is associated with and the higher the triggering values of these associated atoms is, the higher becomes the triggering value of $\mathfrak{a}$, too.

Example 8. *Figure 2 shows on the left-hand side the (unlabeled) spreading activation network of $\Delta^\mathfrak{a}$ (Table 1). The labelings with respect to the queries $\mathfrak{q}_1$ and $\mathfrak{q}_2$ are shown on the right-hand side. For example, $\Sigma(\mathfrak{q}_1) = \{a, p\}$ and consequently $label(a) = label(p) = 1$. Next, b, f, i, and w are labeled as they are direct neighbors of at least one of the atoms a, p. For instance, $\{a, w\} \in \mathcal{E}$ and, therefore,*

$$label(w) = \frac{label(a)}{1 + label(a) + label(p)} = 1/3.$$

> **Labeling Algorithm**
>
> **Input:** Spreading activation network $\mathcal{N}(\Delta) = (\mathcal{V}, \mathcal{E})$ (unlabeled);
> query $\mathfrak{q} = (B|A)$
> **Output:** Labeling of $\mathcal{N}(\Delta)$, i.e., triggering values $\tau_\mathfrak{q}^\Delta(\mathfrak{a}) = label(\mathfrak{a})$ for $\mathfrak{a} \in \Sigma$
>
> 1   for $\mathfrak{a} \in \mathcal{V}$ with $\mathfrak{a} \in \Sigma(\mathfrak{q})$ do
> 2 $label(\mathfrak{a}) = 1$
> 3   initialize
> 4 $\mathcal{L} = \{\mathfrak{a} \in \mathcal{V} \mid \mathfrak{a} \text{ is labeled}\}$,
> 5 $\mathcal{V}' = \{\mathfrak{a} \in \mathcal{V} \mid \exists \{\mathfrak{a}, \mathfrak{b}\} \in \mathcal{E}: \mathfrak{a} \in \mathcal{V} \setminus \mathcal{L} \wedge \mathfrak{b} \in \mathcal{L}\}$
> 6   while $\mathcal{V}' \neq \emptyset$ do
> 7   for $\mathfrak{a} \in \mathcal{V}'$ do
> 8 $label(\mathfrak{a}) = \dfrac{\sum_{\mathfrak{b} \in \mathcal{L}:\ \{\mathfrak{a},\mathfrak{b}\} \in \mathcal{E}} label(\mathfrak{b})}{1 + \sum_{\mathfrak{b} \in \mathcal{L}} label(\mathfrak{b})}$
> 9   update $\mathcal{L}, \mathcal{V}'$
> 10 for $\mathfrak{a} \in \mathcal{V} \setminus \mathcal{L}$ do
> 11 $label(\mathfrak{a}) = 0$
> 12 return $label(\mathfrak{a})$ for $\mathfrak{a} \in \mathcal{V}$

Figure 3: Labeling of a spreading activation network $\mathcal{N}(\Delta)$ wrt. a query $\mathfrak{q}$.

Atom b is neighbor of a and p and is labeled with

$$label(b) = \frac{label(a) + label(p)}{1 + label(a) + label(p)} = 2/3.$$

Eventually, we follow the idea that a conditional $\mathfrak{r}$ cannot be triggered more than the atoms mentioned in the conditional and define the weighting factor by

$$\mathcal{W}_\mathfrak{q}^\Delta(\mathfrak{r}) = \min\{\tau_\mathfrak{q}^\Delta(\mathfrak{a}) \mid \mathfrak{a} \in \Sigma(\mathfrak{r})\}.$$

Example 9. *The weighting factors of the conditionals in $\Delta^\mathfrak{a}$ (Table 1) with respect to the queries $\mathfrak{q}_1$ and $\mathfrak{q}_2$ are shown in Table 3. The weighting factors depend on the labeling of the spreading activation network in Figure 2 which is explained in the next paragraph. For example, $\tau_{\mathfrak{q}_1}^{\Delta^\mathfrak{a}}(c) = 4/15$ and $\tau_{\mathfrak{q}_1}^{\Delta^\mathfrak{a}}(f) = 2/3$. Consequently, the weighting factor of $\mathfrak{r}_9$ with respect to $\mathfrak{q}_1$ is $\mathcal{W}_{\mathfrak{q}_1}^{\Delta^\mathfrak{a}}(\mathfrak{r}_9) = \min\{4/15, 2/3\} = 4/15$.*

$\mathcal{S}(\mathfrak{r}_i,\mathfrak{r}_j)$	$\mathfrak{r}_1$	$\mathfrak{r}_2$	$\mathfrak{r}_3$	$\mathfrak{r}_4$	$\mathfrak{r}_5$	$\mathfrak{r}_6$	$\mathfrak{r}_7$	$\mathfrak{r}_8$	$\mathfrak{r}_9$	$\mathfrak{r}_{10}$	$\mathfrak{r}_{11}$	$\mathfrak{r}_{12}$	$\mathfrak{r}_{13}$	$\mathfrak{r}_{14}$	$\mathfrak{r}_{15}$	$\mathfrak{r}_{16}$	$\mathfrak{r}_{17}$	$\mathfrak{r}_{18}$	$\mathfrak{r}_{19}$	$\mathfrak{r}_{20}$
$\mathfrak{r}_1$	1	1	1/4	1/4			1/4		1/4	1/5		1/4							1/2	
$\mathfrak{r}_2$		1	1/4	1/4			1/4		1/4	1/5		1/4							1/2	
$\mathfrak{r}_3$			1	1/3	1/3	1/3	1/3	1/3	1/3	1/4		1/3							1/4	
$\mathfrak{r}_4$				1	1/3	1/3	1/3	1/3	1/3	1/4										
$\mathfrak{r}_5$					1	1/3		1/3												
$\mathfrak{r}_6$						1	1/3	1/3												
$\mathfrak{r}_7$							1	1/3											1/4	
$\mathfrak{r}_8$								1			1/3									
$\mathfrak{r}_9$									1	2/3	1/3								1/4	
$\mathfrak{r}_{10}$										1	2/3								1/5	
$\mathfrak{r}_{11}$											1		1/3	1/3	1/4	1/3	1/3			
$\mathfrak{r}_{12}$												1	1/3	1/3						
$\mathfrak{r}_{13}$													1	1/3	2/3	1/3	1/3	1/3		
$\mathfrak{r}_{14}$														1	2/3	1/4	1/3	1/4		
$\mathfrak{r}_{15}$															1		1/4	1/3		
$\mathfrak{r}_{16}$																1	1/3	1/3		
$\mathfrak{r}_{17}$																	1	1		
$\mathfrak{r}_{18}$																		1	1/4	
$\mathfrak{r}_{19}$																			1	
$\mathfrak{r}_{20}$																				1

Table 4: Degrees of association $\mathcal{S}(\mathfrak{r}_i,\mathfrak{r}_j)$ between the conditionals $\mathfrak{r}_i,\mathfrak{r}_j \in \Delta^a$. Since $\mathcal{S}(\mathfrak{r}_i,\mathfrak{r}_j)$ is symmetric in its arguments, only the entries in the upper right triangle of the table are shown. Also 0-entries are left out for a better readability.

Degree of Association

The degree of association $\mathcal{S}(\mathfrak{r}_i, \mathfrak{r}_j)$ is a measure of connectedness between the conditionals in Δ and is defined by

$$S(\mathfrak{r}_i, \mathfrak{r}_j) = \frac{|\Sigma(\mathfrak{r}_i) \cap \Sigma(\mathfrak{r}_j)|}{|\Sigma(\mathfrak{r}_i) \cup \Sigma(\mathfrak{r}_j)|}, \qquad \mathfrak{r}_i, \mathfrak{r}_j \in \Delta.$$

Hence, it is the number of shared atoms relative to all atoms in $\mathfrak{r}_i$ or $\mathfrak{r}_j$ and, therefore, non-negative and normalized by 1. The degree of association of a conditional $\mathfrak{r}$ to itself is $\mathcal{S}(\mathfrak{r}, \mathfrak{r}) = 1$ while the degree of association of conditionals which do not share any atoms is 0. The syntactically-driven definition of $\mathcal{S}(\mathfrak{r}_i, \mathfrak{r}_j)$ is motivated by and extends the idea of Syntax Splitting in the sense that syntactical dependencies are not understood as a binary relation but are treated as a graduated quantity.

Not only the quantities $\mathcal{S}(\mathfrak{r}_i, \mathfrak{r}_j)$ for $\mathfrak{r}_j \in \Delta$ themselves are essential for the spreading activation of a conditional $\mathfrak{r}_i$ but also how *many* conditionals $\mathfrak{r}_i$ is associated with. The more a conditional is cross-linked within Δ, the more likely it is that this conditional has a high spreading activation and is selected by the selection function s.

Example 10. *The degrees of association between the conditionals in $\Delta^{\mathfrak{a}}$ (Table 1) are shown in Table 4. For example,*

$$\mathcal{S}(\mathfrak{r}_9, \mathfrak{r}_{10}) = \frac{|\{c,f\} \cap \{c,f,s\}|}{|\{c,f\} \cup \{c,f,s\}|} = \frac{|\{c,f\}|}{|\{c,f,s\}|} = \frac{2}{3}.$$

When taking no account of the base-level activation $\mathcal{B}^{\Delta}(\mathfrak{r})$ and considering the spreading activation $\mathcal{S}_{\mathfrak{q}}^{\Delta}(\mathfrak{r})$ only and when $\theta > 0$ in addition, then the selection function $s_{\mathcal{A}}^{\theta}$ selects conditionals which are syntactically linked to the query $\mathfrak{q}$ only. This is because all remaining conditionals $\mathfrak{r}'$ have a spreading activation of $\mathcal{S}_{\mathfrak{q}}^{\Delta}(\mathfrak{r}') = 0 < \theta$. With increasing syntactically linkage to $\mathfrak{q}$ also the spreading activation increases. In this sense, activation-based conditional inference refines Syntax Splitting by a gradual notion. In addition, when taking the base-level activation into account, activation-based conditional inference allows one to soften the syntactically motivated rejection of conditionals which are not linked to the query if the entrenchment of these conditionals is high enough.

Altogether, we are now able to compute $\mathcal{A}_{\mathfrak{q}}^{\Delta}(\mathfrak{r})$ (without usage history).

Example 11. *Table 3 shows $\mathcal{A}_{\mathfrak{q}_i}^{\Delta^{\mathfrak{a}}}$ (cf. also Table 1) wrt. the queries $\mathfrak{q}_1 = (p \Rightarrow a | \top)$ and $\mathfrak{q}_2 = (\overline{f} | c\overline{s})$. If a threshold $\theta = 2.3$ is used, the conditionals which are selected for activation-based conditional inference are*

$$\Delta_1^{\mathfrak{a}} = (\Delta^{\mathfrak{a}})_{\mathcal{A}_1}^{\theta} = \{\mathfrak{r}_1, \mathfrak{r}_2, \mathfrak{r}_3, \mathfrak{r}_6\},$$

where $\mathcal{A}_1 = \mathcal{A}_{\mathfrak{q}_1}^{\Delta^{\mathfrak{a}}}$, and

$$\Delta_2^{\mathfrak{a}} = (\Delta^{\mathfrak{a}})_{\mathcal{A}_2}^{\theta} = \{\mathfrak{r}_8, \mathfrak{r}_9, \mathfrak{r}_{10}, \mathfrak{r}_{11}\},$$

where $\mathcal{A}_2 = \mathcal{A}_{\mathfrak{q}_2}^{\Delta^{\mathfrak{a}}}$. One has $[\![\mathfrak{q}_1]\!]_{\Delta_1^{\mathfrak{a}}}^{\mathcal{I}^P} = yes$ and $[\![\mathfrak{q}_2]\!]_{\Delta_2^{\mathfrak{a}}}^{\mathcal{I}^P} = no$. That is, both queries can already be decided based on the reduced belief bases $\Delta_1^{\mathfrak{a}}$ and $\Delta_2^{\mathfrak{a}}$ with activation-based conditional inference. Note that $\Delta_1^{\mathfrak{a}}$ and $\Delta_2^{\mathfrak{a}}$ are smaller than the resp. direct foci according to (standard) focused inference (cf. Example 6 for $\mathfrak{q}_2$). The threshold $\theta = 2.3$ is such that in both cases 20 % of the conditionals of Δ are selected for drawing the inference.

In the next section, we make the base-level activation dependent on the history of usage of the conditionals and thereby integrate the concepts of forgetting and remembering into activation-based conditional inference.

4.3 Activation-Based Conditional Inference and Forgetting and Remembering

In ACT-R the base-level activation of a chunk is not constant but decreases over time and increases when the chunk is retrieved, reflecting forgetting and remembering. In order to capture this dynamic view on the base-level activation, we decrease the base-level activation of a conditional when the conditional is not selected for answering a query and increase it otherwise. For this, we introduce a *forgetting factor* $\phi_{\delta,s}(\mathfrak{r})$ which is dependent on the selection function $s = s_{\mathcal{A}}^{\theta}(\mathfrak{r})$ of the current inference task and a parameter $\delta \geq 0$, and which is defined by

$$\phi_{\delta,s}(\mathfrak{r}) = \begin{cases} 1 + \delta & \text{iff } s_{\mathcal{A}}^{\theta}(\mathfrak{r}) = 1 \\ 1 - \delta & \text{otherwise} \end{cases}.$$

After performing the inference task, we update the base-level activation with this forgetting factor and obtain for the updated base-level activation

$$\mathcal{B}_{\delta,s}^{\Delta}(\mathfrak{r}) = \mathcal{B}^{\Delta}(\mathfrak{r}) \cdot \phi_{\delta,s}(\mathfrak{r}).$$

The higher the parameter δ, the greater the impact of the forgetting factor $\phi_{\delta,s}(\mathfrak{r})$ on the base-level activation is. When applying this kind of update to a series of inference requests, the usage history of the conditionals is implemented into the base-level activation implicitly.

The next example shows how this update procedure captures the idea of *forgetting*. Here, we understand forgetting as the process of lowering the base-level

Conditional	$\mathcal{B}^{\Delta^a}(\mathfrak{r}_i)$	$\mathcal{A}^{\Delta^a}_{\mathfrak{q}_2}(\mathfrak{r}_i)$	$\mathcal{A}^{\Delta^a}_{\mathfrak{q}_1}(\mathfrak{r}_i)$	$\mathcal{B}^{\Delta^a}_{\mathfrak{q}_1}(\mathfrak{r}_i)$	$\mathcal{A}^{\Delta^a}_{\mathfrak{q}_1,\mathfrak{q}_2}(\mathfrak{r}_i)$
$\mathfrak{r}_1$	1	2.27	⬚2.36⬚	1.20	⬚2.47⬚
$\mathfrak{r}_2$	1	2.27	⬚2.36⬚	1.20	⬚2.47⬚
$\mathfrak{r}_3$	1	1.66	⬚2.41⬚	1.20	1.86
$\mathfrak{r}_4$	1	1.70	2.13	0.80	1.50
$\mathfrak{r}_5$	1	1.41	1.82	0.80	1.21
$\mathfrak{r}_6$	1	1.60	⬚2.36⬚	1.20	1.80
$\mathfrak{r}_7$	1/2	1.60	1.73	0.40	1.50
$\mathfrak{r}_8$	1	⬚2.43⬚	2.03	0.80	2.23
$\mathfrak{r}_9$	1/2	⬚2.85⬚	1.43	0.40	⬚2.75⬚
$\mathfrak{r}_{10}$	1/3	⬚2.94⬚	1.14	0.27	⬚2.88⬚
$\mathfrak{r}_{11}$	1/2	⬚2.58⬚	0.90	0.40	⬚2.48⬚
$\mathfrak{r}_{12}$	1	1.29	1.78	0.80	1.09
$\mathfrak{r}_{13}$	1	1.12	1.29	0.80	0.92
$\mathfrak{r}_{14}$	1	1.08	1.27	0.80	0.88
$\mathfrak{r}_{15}$	1	1.12	1.29	0.80	0.92
$\mathfrak{r}_{16}$	1	1.11	1.19	0.80	0.91
$\mathfrak{r}_{17}$	1	1.07	1.17	0.80	0.87
$\mathfrak{r}_{18}$	1	1.15	1.18	0.80	0.95
$\mathfrak{r}_{19}$	1/2	1.59	1.39	0.40	1.49
$\mathfrak{r}_{20}$	1	1	1	0.80	0.80

Table 5: Activation function $\mathcal{A}^{\Delta^a}_{\mathfrak{q}_1,\mathfrak{q}_2}$ where the base-level activation was updated by $\phi_{\delta,s}$ with $\delta = 0.2$ and $s^{-1}(1) = (\Delta^a)^\theta_{\mathcal{A}^{\Delta^a}_{\mathfrak{q}_1}}$ beforehand. $\mathcal{A}^{\Delta^a}_{\mathfrak{q}_1}$ and $\mathcal{A}^{\Delta^a}_{\mathfrak{q}_2}$ are recalled for comparison. $\mathcal{B}^{\Delta^a}_{\mathfrak{q}_1}(\mathfrak{r}_i)$ is the base-level activation after querying $\mathfrak{q}_1$. Selected conditionals are boxed (threshold $\theta = 2.3$).

activation of a conditional $\mathfrak{r}$ so far over time (measured in the number of processed inference requests) that the conditional $\mathfrak{r}$ is not considered for answering a query $\mathfrak{q}$ after processing some other inference requests $\mathfrak{q}_1, \ldots, \mathfrak{q}_m$ in which $\mathfrak{r}$ did not play a role, although this conditional $\mathfrak{r}$ would have been selected for answering the same query $\mathfrak{q}$ if $\mathfrak{q}$ would have been asked directly, i.e. before $\mathfrak{q}_1, \ldots, \mathfrak{q}_m$. In other words, we say that $\mathfrak{r}$ is *forgotten* during the processing of the queries $\mathfrak{q}_1, \ldots, \mathfrak{q}_m$ when the update of the base-level activation of $\mathfrak{r}$ caused by $\mathfrak{q}_1, \ldots, \mathfrak{q}_m$ (more precisely, caused by the forgetting factors associated with $\mathfrak{q}_1, \ldots, \mathfrak{q}_m$) *is the reason* why $\mathfrak{r}$ is not selected for answering $\mathfrak{q}$.

Example 12. *Let $\Delta^\mathfrak{a}$ as in Table 1. We compare the activation function $\mathcal{A}_{\mathfrak{q}_2}^{\Delta^\mathfrak{a}}$ with respect to the query $\mathfrak{q}_2 = (f|c\bar{s})$ with the activation function $\mathcal{A}_{\mathfrak{q}_1,\mathfrak{q}_2}^{\Delta^\mathfrak{a}}$ which is obtained by querying $\mathfrak{q}_1$ and updating $\mathcal{B}^{\Delta^\mathfrak{a}}$ with respect to $s^{-1}(1) = (\Delta^\mathfrak{a})_{\mathcal{A}_{\mathfrak{q}_1}^{\Delta^\mathfrak{a}}}^\theta$ first and by querying $\mathfrak{q}_2$ afterwards. While in the first case the conditionals selected for activation-based conditional inference are $\{\mathfrak{r}_8, \mathfrak{r}_9, \mathfrak{r}_{10}, \mathfrak{r}_{11}\}$ (cf. Example 11), in the second case $\{\mathfrak{r}_1, \mathfrak{r}_2, \mathfrak{r}_9, \mathfrak{r}_{10}, \mathfrak{r}_{11}\}$ are selected (cf. Table 5, also for the used parameters; for answering all inference requests, we chose the same threshold $\theta \geq 2.3$). In particular, this means that $\mathfrak{r}_8$ is forgotten because it would have been selected for answering $\mathfrak{q}_2$ if $\mathfrak{q}_2$ would have been asked first, but it is not selected in the case where $\mathfrak{q}_2$ is asked after $\mathfrak{q}_1$. The forgetting of $\mathfrak{r}_8$ happens because $\mathfrak{r}_8$ did not play a role when answering $\mathfrak{q}_1$ and, hence, the base-level activation of $\mathfrak{r}_8$ is lowered (compare columns 2 and 5 of Table 5). In both cases, the query $\mathfrak{q}_2$ is answered with no.*

The final example shows how remembering is realized within our approach. Remembering a conditional here means that this conditional is selected for answering a query although it has not been selected in the previous reasoning task.

Example 13. *When querying $\mathfrak{q}_1 = (p \Rightarrow a|\top)$ from $\Delta^\mathfrak{a}$ (Table 1) with threshold $\theta = 2.3$, the conditional $\mathfrak{r}_{10}$ is not selected (cf. Table 5) and consequently its base-level activation is decreased (also cf. Table 5). After this, it has the lowest base-level activation of all conditionals in $\Delta^\mathfrak{a}$. However, it turns out that this conditional is selected and, hence, remembered when asking for $\mathfrak{q}_2 = (f|c\bar{s})$ afterwards (cf. Example 12).*

Note that also a stronger variant of remembering can be observed in our approach: It can happen that a conditional is not selected for answering a query $\mathfrak{q}$ but over time this conditional is considered in other reasoning tasks such that its base-level activation increases to such an extend that when it comes to the query $\mathfrak{q}$ again this conditional is selected as of now. Although the base-level activation of a conditional may have been decreased over time through the forgetting factor to nearly zero, the conditional can still be selected by a selection of conditionals s if the spreading activation is high enough to compensate the low base-level activation.

5 Conclusions and Future Work

In this paper, we combined conditional reasoning based on ranking semantics and the cognitive architecture ACT-R [5, 4] and, therewith, developed a prototypical model for activation-based conditional inference. More technically, we reformulated the activation function from ACT-R for conditionals and selected the conditionals with the highest degree of activation for focused inference [23]. Focused inference means drawing inferences from a subset of a belief base instead of considering the whole belief base. One motivation for doing so, besides computational aspects, is to mimic human reasoning more adequately. For example, with activation-based conditional inference it is possible to implement several aspects of human reasoning into modern expert systems such as focusing, forgetting, and remembering.

The activation function in ACT-R comprises several cognitive concepts which describe how humans retrieve their beliefs when solving reasoning tasks. Basically, it splits into the base-level activation formalizing the entrenchment of the beliefs in the reasoner's memory and the spreading activation which reflects how strongly a reasoning task, here an inference request, triggers the single beliefs. In this paper, we gave a blueprint on how these cognitive concepts can be formalized in a conditional logical setting and proved its beneficing by means of illustrating examples. Therewith, we detached ACT-R from its production system-based inference engine and developed a test field for cognitive conditional reasoning.

The main challenge for future work is to find for a given query q and a given inference operator $\mathfrak{I}$ a proper least subset Δ' of a belief base Δ such that the query q is answered the same with respect to Δ' as to Δ, i.e., $[\![q]\!]^{\mathfrak{I}}_{\Delta'} = [\![q]\!]^{\mathfrak{I}}_{\Delta}$, without having to draw the computationally expensive inference $[\![q]\!]^{\mathfrak{I}}_{\Delta}$. This would strengthen the justification of focused inference in view of both relevance and computational aspects. A helpful property in this context might be Semi-Monotony which is, for instance, satisfied by the System P inference operator $\mathfrak{I}^P$ and which guarantees that no other inferences can be drawn from subsets of the belief base Δ than from Δ itself. Further tasks for future work are the training of our model on real test data in order to find suitable choices for the parameters of our model, and the investigation of the complexity of our approach. We expect that the benefit of our approach in terms of computational costs heavily depends on the chosen inference formalism and cannot be meaningfully answered in general. At least the (history-independent part of the) base-level activation and the degrees of association are independent of the inference query and can be pre-calculated for each belief base.

Acknowledgments

This work was supported by DFG Grant KE 1413/12-1 awarded to Gabriele Kern-Isberner and DFG Grant BE 1700/10-1 awarded to Christoph Beierle as part of the priority program "Intentional Forgetting in Organizations" (SPP 1921).

References

[1] Adams, E.W.: The Logic of Conditionals. Springer (1975)

[2] Anderson, J.R.: Human Associative Memory: A Brief Edition. Lawrence Erlbaum Assoc. (1980)

[3] Anderson, J.R.: A spreading activation theory of memory. Journal of Verbal Learning and Verbal Behavior **22**, 261–295 (1983)

[4] Anderson, J.R.: How can the human mind occur in the physical universe? Oxford University Press (2007)

[5] Anderson, J.R., Lebiere, C.: The atomic components of thought. Psychology Press (1998)

[6] de Finetti, B.: La prévision, ses lois logiques et ses sources subjectives. Ann. Inst. H. Poincaré **7**(1), 1–68 (1937), English translation in *Studies in Subjective Probability*, ed. H. Kyburg and H.E. Smokler, 1974, 93–158. New York: Wiley & Sons

[7] Goldszmidt, M., Pearl, J.: Qualitative probabilities for default reasoning, belief revision, and causal modeling. Artificial Intelligence **84**, 57–112 (1996)

[8] Goldszmidt, M., Pearl, J.: On the consistency of defeasible databases. Artif. Intell. **52**(2), 121–149 (1991)

[9] Jackson, P.: Introduction to Expert Systems. Addison Wesley, 3rd edn. (1999)

[10] Kern-Isberner, G.: A thorough axiomatization of a principle of conditional preservation in belief revision. Ann. Math. Artif. Intell. **40**(1-2), 127–164 (2004)

[11] Kern-Isberner, G., Beierle, C., Brewka, G.: Syntax splitting = relevance + independence: New postulates for nonmonotonic reasoning from conditional belief bases. In: Calvanese, D., Erdem, E., Thielscher, M. (eds.) Proceedings of the 17th International Conference on Principles of Knowledge Representation and Reasoning, KR 2020. pp. 560–571 (2020)

[12] Klahr, D., Langley, P., Neches, R. (eds.): Production System Models of Learning and Development. MIT Press (1987)

[13] Kraus, S., Lehmann, D., Magidor, M.: Nonmonotonic reasoning, preferential models and cumulative logics. Artif. Intell. **44**(1-2), 167–207 (1990)

[14] Kutsch, S., Beierle, C.: InfOCF-Web: An online tool for nonmonotonic reasoning with conditionals and ranking functions. In: Zhou, Z. (ed.) Proceedings of the Thirtieth International Joint Conference on Artificial Intelligence, IJCAI 2021, Virtual Event / Montreal, Canada, 19-27 August 2021. pp. 4996–4999. ijcai.org (2021)

[15] Lehmann, D., Magidor, M.: What does a conditional knowledge base entail? Artif. Intell. **55**(1), 1–60 (1992)

[16] Newell, A.: Unified Theories of Cognition. Harvard University Press (1990)

[17] Pearl, J.: System Z: A natural ordering of defaults with tractable applications to nonmonotonic reasoning. In: Parikh, R. (ed.) Proceedings of the 3rd Conference on Theoretical Aspects of Reasoning about Knowledge. pp. 121–135. Morgan Kaufmann (1990)

[18] Rajendra, A., Sajja, P.: Knowledge-Based Systems. Jones and Bartlett Learning (2009)

[19] Ryle, G.: The Concept of Mind. University of Chicago Press, New edn. (2000)
[20] Spohn, W.: The Laws of Belief: Ranking Theory and Its Philosophical Applications. Oxford University Press (2012)
[21] Wilhelm, M., Howey, D., Kern-Isberner, G., Sauerwald, K., Beierle, C.: A brief introduction into activation-based conditional inference. In: Beierle, C., Ragni, M., Stolzenburg, F., Thimm, M. (eds.) Proceedings of the 7th Workshop on Formal and Cognitive Reasoning co-located with the 44th German Conference on Artificial Intelligence (KI 2021), September 28, 2021. CEUR Workshop Proceedings, vol. 2961, pp. 4–8. CEUR-WS.org (2021)
[22] Wilhelm, M., Howey, D., Kern-Isberner, G., Sauerwald, K., Beierle, C.: Integrating cognitive principles from ACT-R into probabilistic conditional reasoning by taking the example of maximum entropy reasoning. In: Barták, R., Keshtkar, F., Franklin, M. (eds.) Proceedings of the Thirty-Fifth International Florida Artificial Intelligence Research Society Conference, FLAIRS 2022, Hutchinson Island, Jensen Beach, Florida, USA, May 15-18, 2022 (2022)
[23] Wilhelm, M., Kern-Isberner, G.: Focused inference and System P. In: Thirty-Fifth AAAI Conference on Artificial Intelligence, AAAI 2021. pp. 6522–6529. AAAI Press (2021)

Do Humans Find Postulates of Belief Change Plausible?

Clayton Kevin Baker *
University of Cape Town and CAIR, Cape Town, South Africa
bkrcla003@myuct.ac.za

Thomas Meyer
University of Cape Town and CAIR, Cape Town, South Africa
tmeyer@cs.uct.ac.za

Abstract

Various empirical methods were used to test whether humans agree with postulates of non-monotonic reasoning and belief change. This work investigates through surveys whether postulates of revision and update are plausible with human reasoners when presented as material implication statements. We used statistical methods to measure the association between the antecedent and the consequent of each postulate. The results show that participants tend to find postulates of update more plausible than postulates of revision.

1 Introduction

The study of non-monotonic reasoning and belief change presents a formal way for a reasoner to change their beliefs to accommodate new information. This is similar to how humans reason every day. The hallmark non-monotonic reasoning problem in AI is that of "Tweety the bird" which presents the evidence that penguins are non-flying birds and that this evidence defeats the accepted premise that birds can fly [28]. In the cognitive science and AI communities, an ongoing goal is to model and predict the way people draw conclusions in everyday situations [31]. Ragni et al. [29] argue that future technologies will increasingly interact with humans and demonstrate cognitive features such as tolerance to exceptions, robustness, and the

---

*We wish to express our sincere gratitude and appreciation to the DSI–CSIR Interbursary Support (IBS) Programme and the Centre for Artificial Intelligence Research (CAIR) for financial support.

flexibility to accommodate new information. Human belief change has been studied using various approaches in the AI and cognitive science communities. The studies that stem from AI focus on how human reasoners judge the content and structure of logical arguments, and what can be inferred from a set of facts. Experiments in the form of surveys test how logic-based theories and properties are received by human reasoners. For example, Da Silva Neves et al. [24] surveyed English translations of the postulates of defeasible reasoning, judging the premises and conclusion separately, for plausibility with human reasoners. Ragni et al. [27] compared predictions of logical systems of defeasible reasoning, as opposed to a set of postulates, to the Suppression Task [7], before surveying English translations of the predictions with human reasoners. The studies that stem from cognitive science use experiments to understand the factors that influence cognition and construct mental models that explain human belief change. For example, Knauf et al. [21] study how humans revise their beliefs about objects and its position in space when presented with new information. We make two hypotheses about humans and belief change. We hypothesise that human reasoners judge the postulates for belief revision, given by Alchourrón, Gärdenfors and Makinson (AGM) [1], as true. We also hypothesise that human reasoners judge the postulates for belief update, given by Katsuno and Mendelzon (KM) [20], as true. The postulates are formulated in propositional logic and we test this hypothesis empirically. Additionally, we note that once our hypotheses are tested, our results can be transferred to other forms of logic.

The outline of the rest of this paper is as follows. In Section 2, we give preliminaries for propositional logic and introduce the postulates for revision and update. In Section 3, we describe our choice of representation for the postulates, the experimental setup, and ethical issues. We report on our experimental data and initial results in Section 4. In Section 5, we discuss the results from our experiments and suggest avenues for future work. Furthermore, we note that this work builds on previous work that investigated the relationship between postulates of defeasible reasoning [3] and belief change [4] with human reasoners.

2 Background

2.1 Preliminaries

We consider an infinite propositional language $\mathcal{L}$ and denote the set consisting of all the propositional atoms in $\mathcal{L}$ by Σ. Atoms are represented by lowercase Roman letters e.g. a, b, c, etc., and formulas are denoted by lowercase Greek letters e.g. α, β, γ, etc. In the belief change community, it is typical to represent the beliefs of a reasoner in one of three ways. The first is a knowledge base $\mathcal{K}$ that contains a finite set of beliefs. The second is a belief set that uses a logically closed set of formulas e.g. $\mathcal{K}$ such that $\mathcal{K} = C_n(\mathcal{K})$ where C_n is the logical closure operator. The last is a belief base that uses a formula ψ to represent a reasoner's beliefs. We use the latter two for revision (Section 2.2) and update (Section 2.3), respectively. Boolean operations of negation ($\neg$), conjunction ($\wedge$), disjunction ($\vee$), material implication ($\rightarrow$)and material equivalence ($\leftrightarrow$) are used to combine propositional formulas. In addition, we define the notions of interpretation, satisfaction, and models in the following. An interpretation of $\mathcal{L}$ is a function from Σ to {T,F}. We denote an interpretation by a tuple representing each propositional atom's value, e.g. if $\Sigma = \{e, f, g, h\}$ then <T, F, T, F> is the interpretation which maps e, f, g, h to T, F, T, F respectively. A model of a propositional formula α is an interpretation that makes α true. We say that a propositional formula δ is complete if for any propositional formula, ν, δ implies ν or δ implies $\neg\nu$. A set of propositional formulas is satisfiable if at least one interpretation makes every formula in the set true. The set of interpretations that satisfy a formula α is called the set of models of α, and is denoted by $Mod(\alpha)$.

2.2 Postulates for revision

Alchourrón, Gärdenfors and Makinson (AGM) [1] produced influential work in the study of theory contraction and revision. Contraction is the process of reducing a set of sentences to take out a proposition while revision incorporates a proposition into a set of sentences. They investigated partial meet contraction functions and defined the basic postulates of these functions. They have shown that the properties of partial meet contraction functions, viz. closure, success, inclusion, vacuity, recovery, and extensionality, satisfy the Gärdenfors rationality postulates [12] and that they are sufficiently general to provide a representation theorem for those postulates. An important outcome of their work is the properties and representation theorem for contraction functions, which have been extended to revision functions in later work. Revision is an approach to reasoning with changing beliefs under the assumption that the world did not undergo a fundamental change. A revision operation allows a reasoner to add new information to his beliefs if the new information is consistent with his beliefs. A revision operation also allows a reasoner to add an exception to his beliefs to account for the situation where this exception or new information is inconsistent with his beliefs. Moreover, the result of a revision operation must always be that a reasoner's beliefs do not contradict one another. Katsuno and Mendelzon [19] investigated the semantics of revising a belief base with sets of propositions. In the KM framework for set revision, $\mathcal{K}$ is a deductively closed belief set and μ and ϕ are formulas. $\mathcal{K} * \mu$ means the revision of $\mathcal{K}$ by μ. $\mathcal{K} + \mu$ is the smallest deductively closed set containing $\mathcal{K}$ and μ. $\mathcal{K}_\perp$ is the set consisting of all the propositional formulas. The KM postulates for set revision are given by (∗1)–(∗8).

(∗ 1) $\mathcal{K} * \mu$ is a belief set.

(∗ 2) $\mu \in K * \mu$

(∗ 3) $\mathcal{K} * \mu \subseteq \mathcal{K} + \mu$

(∗ 4) If $\neg \mu \notin \mathcal{K}$, then $\mathcal{K} + \mu \subseteq \mathcal{K} * \mu$

(∗ 5) $K * \mu = \mathcal{K}_\perp$ only if μ is unsatisfiable

(∗ 6) If $\mu \equiv \phi$ then $\mathcal{K} * \mu \equiv \mathcal{K} * \phi$.

(∗ 7) $\mathcal{K} * (\mu \wedge \phi) \subseteq (\mathcal{K} * \mu) + \phi$

(∗ 8) If $\neg \phi \notin \mathcal{K} * \mu$ then $(\mathcal{K} * \mu) + \phi \subseteq \mathcal{K} * (\mu \wedge \phi)$

2.3 Postulates for update

Katsuno and Mendelzon [18] gave a characterisation of all revision methods that satisfy the AGM postulates in terms of a pre-order among models. In subsequent work [20], they defined postulates for updating a finite propositional knowledge base by partial orders or partial pre-orders over interpretations. The class of operators defined generalises Winslett's Possible Models Approach (PMA) [34], which Katsuno and Mendelzon argue is an update operator given that the PMA changes each world independently. Herzig and Rifi [17] used the KM postulates for update as a standard for evaluating ten different propositional update operations in the literature. In later work, Herzig et al. [16] studied a family of belief update operators by analysing the interplay between formulas and literals. They defined the operation of update as follows: first, omit from the belief base every literal on which the input formula has a negative impact and then conjoin the resulting base with the input formula. They evaluated the update operators in two dimensions: the logical dimension, by checking the status of KM postulates, and the computational dimension, by identifying the complexity of several decision problems. The KM postulates have also been used by Miller and Muise [23] to evaluate a belief update mechanism for Proper Epistemic Knowledge Bases. This mechanism guarantees consistency in the knowledge base when new beliefs are added. More recently, Creignou et al. [9] argued that belief update within fragments of classical logic has not been addressed thus far. They investigated the behaviour of refined update operators concerning the satisfaction of the KM postulates and, in this context, highlighted the differences between revision and update. Ribeiro et al. [30] used the KM postulates to define a class of belief update functions, called royal splinter functions, for non-finitary logics. Update is an approach to reasoning with changing beliefs after some fundamental shift in the world occurred. The KM framework for update uses a formula ψ to denote a belief base. When we update ψ with new information μ, written $\psi \diamond \mu$, we are saying that we used to believe ψ, we know now that μ holds, and we need to modify ψ by adding μ, acknowledging that we may have been wrong if μ contradicts ψ. The KM postulates for base update are given by $(\diamond 1)$–$(\diamond 8)$:

($\diamond$ 1) $\psi \diamond \mu$ implies μ

($\diamond$ 2) If ψ implies μ then $\psi \diamond \mu$ is equivalent to ψ

($\diamond$ 3) If both ψ and μ are satisfiable then $\psi \diamond \mu$ is also satisfiable

($\diamond$ 4) If $\psi_1 \leftrightarrow \psi_2$ and $\mu_1 \leftrightarrow \mu_2$ then $\psi_1 \diamond \mu_1 \leftrightarrow \psi_2 \diamond \mu_2$

($\diamond$ 5) $(\psi \diamond \mu) \wedge \phi$ implies $\psi \diamond (\mu \wedge \phi)$

($\diamond$ 6) If $\psi \diamond \mu_1$ implies μ_2 and $\psi \diamond \mu_2$ implies μ_1 then $\psi \diamond \mu_1 \leftrightarrow \psi \diamond \mu_2$

($\diamond$ 7) If ψ is complete then $(\psi \diamond \mu_1) \wedge (\psi \diamond \mu_2)$ implies $\psi \diamond (\mu_1 \vee \mu_2)$

($\diamond$ 8) $(\psi_1 \vee \psi_2) \diamond \mu \leftrightarrow (\psi_1 \diamond \mu) \vee (\psi_2 \diamond \mu)$

($\diamond$ 9) If ψ is complete and $(\psi \diamond \mu) \wedge \phi$ is satisfiable then $\psi \diamond (\mu \wedge \phi)$ implies $(\psi \diamond \mu) \wedge \phi$

Revision and update differ from non-monotonic logic using the concept of orders on interpretations. A homogeneous relation $\leq$ on some given set P, so that by definition $\leq$ is some subset of $P \times P$ and the notation $a \leq b$ is used in place of $(a, b) \in P$, is called a preorder if the relation is also transitive and reflexive. A reflexive relation has the property that $a \leq a$ for all $a \in P$. A transitive relation has the property that if $a \leq b$ and $b \leq c$ then $a \leq c$ for all $a, b, c \in P$. If a preorder is also anti-symmetric, that is, $a \leq b$ and $b \leq a$ implies $a = b$, then it is a partial preorder. A preorder is total if $a \leq b$ or $b \leq a$ for all $a, b \in P$. A revision operator satisfies postulates ($*$ 1)–($*$ 8) using the notion of a total preorder on interpretations while an update operator satisfies postulates ($\diamond$ 1)–($\diamond$ 6) using partial preorders on interpretations. Katsuno and Mendelzon [20] have shown that by replacing postulates ($\diamond$ 6) and ($\diamond$ 7) with a new postulate ($\diamond$ 9), the class of update operators can be designed using total preorders. The second and more important difference between revision and update is that, in the case of update, a different ordering is induced by each model of ψ, while for revision, only one ordering is induced by the whole of $\mathcal{K}$.

3 Methodology

The methodology we employ uses empirical evidence and statistical analysis to determine whether our hypotheses are true. We present the postulates as material implication statements and translate them into English before surveying them with human reasoners for plausibility. The survey material is given in Appendix B (Experiment 1) and Appendix C (Experiment 2). The symbol $\top$ denotes a tautological antecedent. The single turnstile $\vdash$ denotes entailment e.g. $\alpha \vdash \gamma$ is taken as "from α it follows that γ holds". For consistent notation amongst both sets of postulates, some symbols and postulate metadata have been changed. The postulate metadata "is equivalent to" was replaced with $\equiv$, and "implies" was replaced with $\vdash$. The symbol $\leftrightarrow$ was replaced with $\equiv$. The antecedent and consequent for each postulate was translated into English and surveyed with human reasoners for plausibility on Amazon Mechanical Turk (MTurk). In our analysis, we discuss the effect of the antecedent and the consequent on the plausibility of each postulate.

3.1 Methods of analysis

We compute the number of judgements, given on a scale from 1 (not plausible) to 10 (plausible) of the antecedent and the consequent of each postulate. The variables of our study are the judgement of the antecedent and the consequent, respectively. The data is captured in a 2 × 2 contingency table. We assume the relationship between the variables is linear. The phi-coefficient [15] is used to measure the strength of the relation. The phi-coefficient formula for the 2 × 2 contingency table is:

$$\phi = \frac{TT.FF \quad TF.FT}{\sqrt{((TT+TF)(FT+FF)(TT+FT)(TF+FF))}} \quad (1)$$

The values of T and F in Equation 1 correspond to a combination of the number of judgements of the antecedent and consequent. For example, TT is the number of judgements of the antecedent and the consequent being true. The phi-coefficient produces a value in {-1;1}. Two independent variables are strongly correlated when ϕ is close to 1, and weakly correlated when ϕ is close to 0. A negative value of ϕ indicates an inverse relationship exists between two independent variables. A value of 0 means there is no relationship. An error can occur when the denominator of the phi-coefficient evaluates to 0, because of any of the four sums in the denominator evaluating to 0. This is treated by setting the denominator to 1, an arbitrary value [32], which means that no association exists. We compute the significance of the phi-coefficient using Fisher's exact test [14] to test whether the relation holds for our participants and given our relatively small sample size of fewer than 60 participants.

3.2 Ethical issues

We obtained ethical clearance from the Faculty of Science Ethics Research Committee at the University of Cape Town. We include the consent forms and a link to our data management plan in our GitHub project repository, linked in Appendix A. For the design of our experiments, we utilised Google Forms to create surveys. The surveys were hosted on MTurk from which we crowdsourced our data collection. Google Forms, Qualtrics, and SurveyMonkey are examples of tools for creating online surveys with a variety of question types. The surveys are typically distributed by email invitation or by sharing a unique URL. In terms of data collection for these tools, the onus is on the survey creator to identify participants and elicit responses. In contrast, crowdsourcing is a means of gathering a response to a task from a large audience, usually via the internet. MTurk builds on the crowdsourcing concept by facilitating task creation, hosting tasks, and managing data collection through its database of registered participants. Bentley et al. [5] found that crowdsourced surveys are completed faster and cost less than traditional surveys, with no significant ($< 10\%$) loss in accuracy. To ensure that responses are of acceptable quality, Grootswagers [13] suggests the inclusion of a trial task before the primary task to give participants an idea of what is expected from them. An issue with anonymous data collection from MTurk is that unsupervised participants tend to be less attentive than supervised participants. This is mitigated by checking in on participants at various stages of the task. The check-in aims to remind inattentive participants to pay more attention, typically through asking a trial question or manipulating an instruction [25]. When check-ins are conducted this way, there is less experimenter bias [26], subject crosstalk [10], and participant reactance. Springer et al. [33] recommend using a neutral non-persuasive tone in the language of the task to diminish the likelihood of selection bias based on persuasive language and target participant traits. The effect of participant performance on MTurk at different times of the day is robust, while there is some variation in participants' personality and prior experience across these recruitment times [2]. Another concern of mass anonymised data collection is accepting participant responses at face value. This is problematic because participants who depend on MTurk as a source of income are primarily incentivised by monetary reward rather than interest in the task content. Participants also learn and apply behaviour from similar tasks which introduce unknown bias in their responses. MTurk has been shown to have limited ideological representation, in particular, where subjects hold more liberal attitudes than the general public [6]. The difference in convenience and probability sampling should be well-substantiated by repeat sampling over time rather than an analysis of background characteristics alone [8]. We also used Mechanical Turk in previous work [3] to survey the correspondence of human reasoning with defeasible reasoning, belief revision, and belief update respectively. We recorded a generally positive experience with a survey turnaround time below an hour for each participant and compensated Workers commensurate with the local minimum hourly wage.

4 Experiments

We report on the design, participants, and observations for two experiments used to test the plausibility of the revision and update postulates, respectively, with human reasoners.

4.1 Experiment 1

35 MTurk participants (13 female and 22 male) judged the plausibility of English translations of the KM revision postulates $(*1)$–$(*8)$. The task resembled a survey. The antecedent and consequent of each postulate was presented separately. The responses were indicated on a scale from 1 (extremely implausible) to 10 (extremely plausible). Participant responses were translated from numbers on a scale to binary

Postulate	$p \to q$						
	TT	TF	FT	FF	ϕ	Fisher's test	p-value
$(*1)$	31	4	0	0	0	1	not significant
	34	1	0	0	0	1	not significant
$(*2)$	30	5	0	0	0	1	not significant
$(*3)$	31	1	2	1	0.36	0.1664	not significant
$(*4)$	24	4	7	0	0.18	0.562	not significant
$(*5)$	32	1	0	2	0.8	0.0053	significant
$(*6)$	4	9	5	17	0.09	0.6978	not significant
$(*7)$	31	4	0	0	0	1	not significant
$(*8)$	19	11	0	5	0.44	0.0135	significant

Table 1: Participant judgements of the AGM postulates

values. We used the midpoint of the scale to demarcate judgements as true (6 to 10) and false (1 to 5). In the case of a postulate antecedent comprising multiple statements, the final judgement (true or false) of the antecedent was taken as the logical value of the conjunction of all the statements in the antecedent. Postulate consequents comprising multiple statements were treated in the same way. In Table

1, TT refers to the number of participants who judged both the antecedent and the consequent of a postulate as true. TF refers to the number of participants who judged the antecedent as true, but the consequent as false. FT and FF are the inverse judgements of TF and TT. Postulates ($*$ 1) and ($*$ 2) have FT and FF arbitrarily set to 0 to account for our assumption that the antecedent of these postulates is always true. The phi-coefficient was used to measure the relation between the judgement of the antecedent and the consequent of each postulate. For postulate ($*$ 5), the relationship is strong and positive. For postulate ($*$ 4), the relationship is weak and negative. For postulate ($*$ 3), ($*$ 6), and ($*$ 8), the relationship is weak and positive. For postulates ($*$ 1), ($*$ 2) and ($*$ 7), no relation exists. Fisher's exact test of independence was performed to examine the significance of the linear relation between postulate antecedent and consequent. The relation for postulate ($*$ 5) is significant, which means that participants who judged the antecedent of postulate ($*$ 5) as true tended to judge its consequent as true too. The relation for postulate ($*$ 8) is also significant. However, the number of participants (TF) who violated the material implication rule for postulate ($*$ 8) weakened the trend that the participants who judged the antecedent of postulate ($*$ 8) as true tended to judge its consequent as true too. The results for all other postulates are not significant. Overall, 2 out of 8 postulates were found true by participants when judging the antecedent and the consequent of each postulate separately. These results suggest that the participants do not tend to find the KM revision postulates true.

4.2 Experiment 2

37 MTurk participants (18 female and 17 male) judged the plausibility of English translations of the KM update postulates ($\diamond$ 1)–($\diamond$ 8). The task format was the same as the previous experiment. We applied the same data transformation process as the previous experiment. Postulates ($\diamond$ 1), ($\diamond$ 5), and ($\diamond$ 8) have FT and FF arbitrarily set to 0 to account for our assumption that the antecedent of these postulates is always true. As before, the phi-coefficient was used to measure the relation between the judgement of the antecedent and the consequent of each postulate. For postulate ($\diamond$ 2), ($\diamond$ 4), and ($\diamond$ 6), the relationship is strong and positive. For postulate ($\diamond$ 3), ($\diamond$ 7), and ($\diamond$ 9), the relationship is weak and positive. No relation exists for postulate ($\diamond$ 1), ($\diamond$ 5) and ($\diamond$ 8). According to Fisher's exact test, the relationship for postulate ($\diamond$ 2), ($\diamond$ 3), ($\diamond$ 4), and ($\diamond$ 6)is significant, which means that participants who judged the antecedent of these postulates as true tended to judge its consequent as true too. The results for all other postulates are not significant. Overall, 4 out of 9 postulates were found true by participants when judging the antecedent and the consequent of each postulate separately. These results suggest that the participants tend to find

Postulate	$p \to q$						
	TT	TF	FT	FF	ϕ	Fisher's test	p-value
($\diamond$ 1)	29	8	0	0	0	1	not significant
($\diamond$ 2)	25	1	4	7	0.66	0.0002	significant
($\diamond$ 3)	30	0	5	2	0.49	0.0315	significant
($\diamond$ 4)	13	0	7	17	0.68	0	significant
($\diamond$ 5)	35	2	0	0	0	1	not significant
($\diamond$ 6)	24	5	0	8	0.71	0	significant
($\diamond$ 7)	21	4	7	5	0.28	0.108	not significant
($\diamond$ 8)	20	17	0	0	0	1	not significant
($\diamond$ 9)	24	5	4	4	0.31	0.0784	not significant

Table 2: Participant judgements of the KM postulates

the KM postulates true.

5 Discussion and Conclusions

In the cognitive science and AI communities, an ongoing goal is to explore the correspondence between formal logic, philosophy, and human reasoning. For example, Elio and Pelletier [11] conducted experiments to determine which set of sentences to believe when additional information contradicts the initial set, a problem of belief revision. In parallel, in the AI community, an ongoing goal is to formalise models of human reasoning for use in AI-enriched systems. Ragni et al. [27, 29] have demonstrated that classical logic fails to capture human inference, whereas non-monotonic logic has the potential to do so.

Our work builds on previous empirical studies in which humans judged the postulates of revision and update. The methodology we used to assess participants' judgements of the postulates is refined from Da Silva Neves et al. [24], with the addition of reproducible rigorous statistical analysis. In our experiments, we first presented participant judgements of the revision and update postulates, and tested whether the judgements hold in general for our participants. Overall, 2 out of 8 KM revision postulates were found true by participants when judging the antecedent and the consequent of each postulate separately. These results suggest that the participants do not tend to find the KM revision postulates true. By presenting alternative conclusions to the postulates, we can determine whether the postulate content had an effect on the participants' judgements. In contrast, 4 out of 9 KM update postulates were found true by participants, with higher measures of association than for the revision postulates, using the same material. This suggests that the postulate content was sufficient for participants to understand the KM update postulates. Given that the belief base representation was used for the update postulates, we propose to use the same representation for the revision postulates in a future investigation. With common representations and material, and the potential for alternative conclusions, we can more accurately determine whether human reasoners find postulates of revision and update true.

In future work, we will use a combined theoretical and empirical approach to test the relation between human reasoners and postulates of belief change. The theoretical part will build on this work by exploring inter-postulate and inter-framework relationships. Inter-postulate relationships refer to the postulates that depend on each other or that involve similar underlying conditions. Inter-framework relationships refer to corresponding postulates in the revision and update settings, as well as between other forms of logic. Once these relationships have been identified, they will provide the basis for a refined empirical investigation. The empirical part will focus on identifying representations of the postulates that support both the theory and the beliefs of human reasoners. Since the linear relation between antecedent and consequent is generally weak, we propose to test alternative conclusions to the postulates conclusions with human reasoners. The data from the new investigation will give us a stronger position to determine whether human reasoners find postulates of revision and update plausible. Our analysis will include an evaluation of the new data on Cognitive Computation for Behavioural Reasoning Analysis (CCOBRA)[1], a program that uses cognitive models to predict individual responses. The evaluation will allow us to measure the relation between revision and update for each participant. Furthermore, non-parameterised statistical methods, that is, methods that do not assume how the sample data is distributed, e.g. the Wilcoxon signed-rank test [22], will be used to interpret the significance of the results.

References

[1] ALCHOURRÓN, C. E., GÄRDENFORS, P., AND MAKINSON, D. On the logic of theory change: Partial meet contraction and revision functions. *Journal of Symbolic Logic 50* (1985), 510–530.

[2] ARECHAR, A., KRAFT-TODD, G., AND RAND, D. Turking overtime: how participant characteristics and behavior vary over time and day on amazon mechanical turk. *Journal of the Economic Science Association 3*, 1 (2017), 1–11.

---

[1] https://orca.informatai.uni-freiburg.de/ccobra/

[3] BAKER, C., DENNY, C., FREUND, P., AND MEYER, T. Cognitive defeasible reasoning: the extent to which forms of defeasible reasoning correspond with human reasoning. In *Proceedings of the First Southern African Conference for Artificial Intelligence Research (SACAIR 2020)* (2020), CCIS, Springer, pp. 119–219.

[4] BAKER, C., AND MEYER, T. Belief change in human reasoning: An empirical investigation on mturk. In *Proceedings of the Second Southern African Conference for Artificial Intelligence Research (SACAIR 2021)* (2021), pp. 520–536.

[5] BENTLEY, F., DASKALOVA, N., AND WHITE, B. Comparing the reliability of amazon mechanical turk and survey monkey to traditional market research surveys. In *Proceedings of the 2017 CHI Conference Extended Abstracts on Human Factors in Computing Systems* (2017), pp. 1092–1099.

[6] BERINSKY, A., HUBER, G., AND LENZ, G. Evaluating online labor markets for experimental research: Amazon.com's mechanical turk. *Political Analysis 20*, 3 (2012), 351–368.

[7] BYRNE, R. M. Suppressing valid inferences with conditionals. *Cognition 31*, 1 (1989), 61–83.

[8] COPPOCK, A. Generalizing from survey experiments conducted on mechanical turk: A replication approach. *Political Science Research and Methods 7*, 3 (2019), 613–628.

[9] CREIGNOU, N., KTARI, R., AND PAPINI, O. Belief update within propositional fragments. *Journal of Artificial Intelligence Research 61* (2018), 807–834.

[10] EDLUND, J., SAGARIN, B., SKOWRONSKI, J., JOHNSON, S., AND KUTTER, K. Whatever happens in the laboratory stays in the laboratory: The prevalence and prevention of participant crosstalk. *Personality and Social Psychology Bulletin 35*, 5 (2009), 635–642.

[11] ELIO, R., AND PELLETIER, F. Belief change as propositional update. *Cognitive Science 21*, 4 (1997), 419–460.

[12] GÄRDENFORS, P., AND MAKINSON, D. Revisions of knowledge systems using epistemic entrenchment. In *Proceedings of the 2nd conference on Theoretical aspects of reasoning about knowledge* (1988), pp. 83–95.

[13] GROOTSWAGERS, T. A primer on running human behavioural experiments online. *Behavior research methods* (2020), 1–4.

[14] HASSEMER, J., AND WINTER, B. Decoding gestural iconicity. *Cognitive Science 42*, 8 (2018), 3034–3049.

[15] HATTORI, M., AND OAKSFORD, M. Adaptive non-interventional heuristics for covariation detection in causal induction: Model comparison and rational analysis. *Cognitive science 31*, 5 (2007), 765–814.

[16] HERZIG, A., LANG, J., AND MARQUIS, P. Propositional update operators based on formula/literal dependence. *ACM Transactions on Computational Logic 14*, 3 (2013), 1–31.

[17] HERZIG, A., AND RIFI, O. Propositional belief base update and minimal change. *Artificial Intelligence 115*, 1 (1999), 107–138.

[18] KATSUNO, H., AND MENDELZON, A. A unified view of propositional knowledge base updates. *Artificial Intelligence 11*, 2 (1989), 1413–1419.

[19] KATSUNO, H., AND MENDELZON, A. Propositional knowledge base revision and minimal change. *Artificial Intelligence 3*, 52 (1991), 263–294.

[20] KATSUNO, H., AND MENDELZON, A. O. On the difference between updating a knowledge base and revising it. *Belief revision* (1991), 183.

[21] KNAUFF, M., BUCHER, L., KRUMNACK, A., AND NEJASMIC, J. Spatial belief revision. *Journal of Cognitive Psychology 25*, 2 (2013), 147–156.

[22] LI, X., WU, Y., WEI, M., GUO, Y., YU, Z., WANG, H., LI, Z., AND FAN, H. A novel index of functional connectivity: phase lag based on wilcoxon signed rank test. *Cognitive Neurodynamics 15*, 4 (2021), 621–636.

[23] MILLER, T. AND MUISE, C. Belief update for proper epistemic knowledge bases. In *Proceedings of the Twenty-Fifth International Joint Conference on Artificial Intelligence* (New York, New York, USA, 2016), IJCAI'16, AAAI Press, pp. 1209–1215.

[24] NEVES, R., BONNEFON, J., AND RAUFASTE, E. An empirical test of patterns for nonmonotonic inference. *Annals of Mathematics and Artificial Intelligence 34*, 1-3 (2002), 107–130.

[25] OPPENHEIMER, D., MEYVIS, T., AND DAVIDENKO, N. Instructional manipulation checks: Detecting satisficing to increase statistical power. *Journal of Experimental Social Psychology 45* (2009), 867–872.

[26] ORNE, M. On the social psychology of the psychological experiment: With particular reference to demand characteristics and their implications. *American psychologist 17*, 11 (1962), 776.

[27] RAGNI, M., EICHHORN, C., BOCK, T., KERN-ISBERNER, G., AND TSE, A. Formal nonmonotonic theories and properties of human defeasible reasoning. *Minds and Machines 27* (2017), 79–117.

[28] RAGNI, M., EICHHORN, C., AND KERN-ISBERNER, G. Simulating human inferences in light of new information: A formal analysis. In *Proceedings of the Twenty-Fifth International Joint Conference on Artificial Intelligence (IJCAI 16)* (2016), S. Kambhampati, Ed., IJCAI Press, pp. 2604–2610.

[29] RAGNI, M., KERN-ISBERNER, G., BEIERLE, C., AND SAUERWALD, K. Cognitive logics–features, formalisms, and challenges. In *ECAI 2020*. IOS Press, 2020, pp. 2931–2932.

[30] RIBEIRO, J., NAYAK, A., AND WASSERMANN, R. Belief update without compactness in non-finitary languages. In *Proceedings of the Twenty-Eighth International Joint Conference on Artificial Intelligence (IJCAI-19)* (2019), pp. 1858–1864.

[31] SCHON, C., SIEBERT, S., AND STOLZENBURG, F. The corg project: Cognitive reasoning. *KI-Künstliche Intelligenz 33*, 3 (2019), 293–299.

[32] SOUZA, E., AND NEGRI, T. First prospects in a new approach for structure monitoring from gps multipath effect and wavelet spectrum. *Advances in Space Research 59*, 10 (2017), 2536–2547.

[33] SPRINGER, V., MARTINI, P., LINDSEY, S., AND VEZICH, I. Practice-based considerations for using multi-stage survey design to reach special populations on amazon's mechanical turk. *Survey Practice 9*, 5 (2016), 1–8.

[34] WINSLETT, M. Reasoning about action using a possible models approach. In *Proceedings of the Seventh National Conference on Artificial Intelligence (AAAI 88)* (1988), AAAI Press, pp. 89–93.

A Supplementary Information

The Github repository for this work can be accessed via this URL, https://tinyurl.com/mre8dt84.

B Material for the revision postulates

The material for the revision experiment is given in Table 3–10.

Postulate	Antecedent	Consequent
(∗ 1)		
	⊤	$\mathcal{K} = C_n(\mathcal{K})$
	*assumed to be true	You accept both your beliefs and the consequences of your beliefs.
	⊤	$\mathcal{K} \ast \mu = C_n(\mathcal{K} \ast \mu)$
	*assumed to be true	Changing your beliefs to accept the information that Zeeta M is a classical pianist means the same the consequences of changing your beliefs to accept that Zeeta M is a classical pianist.

Table 3: Material to test (∗ 1)

Postulate	Antecedent	Consequent
(∗ 2)	⊤	$\mu \in \mathcal{K} * \mu$
	*assumed to be true	The information that Chris P is a waiter is a part of your beliefs after changing your beliefs to accept that Chris P is a waiter.

Table 4: Material to test (∗ 2)

Postulate	Antecedent	Consequent
(∗ 3)	$\mathcal{K} * \mu$	$\mathcal{K} + \mu$
	The information that Jacob B drives at night follows from changing your beliefs to accept the information that Jacob B is a truck driver.	The information that Jacob B drives at night follows from adding the information that Jacob B is a truck driver to your beliefs.

Table 5: Material to test (∗ 3)

Postulate	Antecedent	Consequent
(∗ 4)	$\mathcal{K} \not\models \neg\mu$	$\mathcal{K} + \mu \models \phi$
	The information that Jessica B is a yoga instructor does not contradict your beliefs.	The information that Jessica B does teach breathing exercises follows from adding the information that Jessica B is a yoga instructor to your beliefs.
		$\mathcal{K} * \mu \models \phi$
		The information that Jessica B does teach breathing exercises follows from changing your beliefs to accept the information that Jessica B is a yoga instructor.

Table 6: Material to test (∗ 4)

Postulate	Antecedent	Consequent
(∗ 5)	$\mu \not\models \bot$	$\mathcal{K} * \mu \not\models \bot$
	A contradiction does not follow from the information that Wilma D is a car owner.	A contradiction does not follow from changing your beliefs to accept that Wilma D is a car owner.

Table 7: Material to test (∗ 5)

Postulate	Antecedent	Consequent
(∗ 6)	$\mu \equiv \phi$	$\mathcal{K} \ast \mu \models \gamma$
	The information that if Noel W is a firefighter then Noel W is strong is equivalent to the information that either Noel W is not a firefighter or Noel W is strong.	The information that Noel W saves lives follows from changing your beliefs to accept that if Noel W is a firefighter then Noel W is strong.
		$\mathcal{K} \ast \phi \models \gamma$
		The information that Noel W saves lives follows from changing your beliefs to accept that either Noel W is not a firefighter or Noel W is strong.

Table 8: Material to test (∗ 6)

Postulate	Antecedent	Consequent
(∗ 7)		
	$\mathcal{K} * (\mu \wedge \phi) \models \gamma$	$(\mathcal{K} * \mu) + \phi \models \gamma$
	The information that Philip P does carry a gun follows from changing your beliefs to accept both the information Philip P is a police officer and Philip P can arrest a criminal.	The information that Philip P does carry a gun follows from first changing your beliefs to accept that Philip P is a police officer and then changing your beliefs to accept that Philip P can arrest a criminal.

Table 9: Material to test (∗ 7)

Postulate	Antecedent	Consequent
(∗ 8)	$\mathcal{K} * \mu \not\models \neg \phi$	$(\mathcal{K} * \mu) + \phi \models \gamma$
	Changing your beliefs to accept the information that Mark M is a science professor does not contradict the information that Mark M does enjoy solving problems.	The information that Mark M is a good teacher follows from first changing your beliefs to accept the information that Mark M is a science professor, and then adding to your knowledge the information that Mark M does enjoy solving problems.
		$\mathcal{K} * (\mu \wedge \phi) \models \gamma$
		The information that Mark M is a good teacher follows from changing your beliefs to accept both Mark M is a science professor and Mark M does enjoy solving problems.

Table 10: Material to test (∗ 8)

www.ingramcontent.com/pod-product-compliance
Lightning Source LLC
Chambersburg PA
CBHW080451170426
43196CB00016B/2764